Praise for *Four Seas...*

"Vivid and compelling. . . . The experience of growing, harvesting, selling and tasting fruit becomes the central metaphor for a kind of philosophy of sense-celebration, which, like much of Walt Whitman's poetry, becomes a catalogue of affirmation. . . . 'Simple Pleasures,' Masumoto might well have called this book."

—*Washington Post Book World*

"David Mas Masumoto's gentle and occasionally angry ode to his peaches reminds us that great food does not come from trends or hot new chefs, but from elemental things such as a farmer's love for the land, human dignity, pride in craftsmanship, and a sense of family. It is somehow reassuring to read this lyric little book."

—Mark Kurlansky, author of *Cod* and *Salt*

"Masumoto, as with Dillard and Thoreau, has a message for us about living differently, living better, paying close attention to nature, knowing our neighbors and valuing the process as much as the product. . . . Agriculture chews up dilettantes and spits them out. The mere survival of Masumoto's farm . . . argues for his seriousness."

—*Los Angeles Times Book Review*

"Masumoto's books are pensive, lyrical."

—*Seattle Post Intelligencer*

"Without romanticizing farmwork, Masumoto paints a portrait of the relationships needed to produce a ripe peach and deliver it to a super-market 3000 miles away. Even as he inveighs against all the economic disincentives to grow foods with taste, Mas still finds beauty, purpose, and connection in the daily task of raising peaches. We are left aching for the sweetness of real work but knowing its harsh side as well. A peach will never slurp the same."　　　　　　　—Peter Hoffman, Savoy Restaurant; national chair of Chefs Collaborative

"[Masumoto's] graceful storytelling and easy authorial voice make the book an agreeable reader for anyone who enjoys food and quality of life. . . . [He] writes with a fine lyrical hand."

—*The Bloomsbury Review*

"Because I too farm my small vegetable plot organically, I am always taken with Masumoto's struggle to maintain his hard-won crops organically. But I found the most touching part of his memoir to be the account of his parents' sudden evacuation, post–Pearl Harbor, to a concentration camp in the Arizona desert. We need such reminders especially today."　　　　—Maxine Kumin, author of *The Long Marriage*

"Masumoto brings me to tears just describing a walk through his orchard. What does he treasure most? The peach, his family memories, or the written word. He treats all three with eloquent compassion. Profound stuff!"　　　　　—Jeff Smith, The Frugal Gourmet

"[Masumoto's] voice [is] warm and compassionate. . . . Masumoto's gifts as a writer—his thoughtful and genuine nature, his ability to recreate intricate experiences—come through equally well in his section on sound. . . . The family relationships Masumoto cherishes and the connections he draws over time, through generations, give the book an emotional weight it might not have had otherwise. They give it greater significance. They give us a compelling reason to read and to care."　　　　　　　　　　　—*Oakland Tribune*

"If you enjoy good stories, family life, getting the work done, and great, succulent food, *Four Seasons in Five Senses* is the book for you.

David Mas Masumoto is our poet laureate of peaches, the shock of recognition after that first delicious bite, as sweet juice runs down your chin. This is a wonderful book, delectation personified. Enjoy!"
—William Kittredge, author of *The Nature of Generosity* and *Southwestern Homelands*

"Masumoto has become an eloquent voice for that increasingly rare breed, the family farmer. . . . Masumoto is particularly adept at conveying the junction at which tradition and modernity meet . . . enchanting. . . . Intense, sensuous, lyrical, shaped by the sensibility of a poet and the eye of a farmer."
—*Kirkus Reviews*

"Exquisite . . . a visceral journey." —*Elliot Bay Booknotes*

"When is the story of a peach more than just a fruity tale? When it's told by a farmer with pure poetic passion. . . . Masumoto has a way with words, as he does with fruit." —*Mercury News* (San Jose)

"Masumoto paints pictures with the simple words of a farmer. . . . He retains the dedication of a true artist."
—California Rare Fruit Growers, Inc.

"Take away [Masumoto's] overalls and work boots and you find a poet and a philosopher. . . . Eating this summer's first peach will have new meaning!" —*Publishers Weekly Daily*

FOUR SEASONS

in

FIVE SENSES

Things Worth Savoring

By David Mas Masumoto

W. W. NORTON & COMPANY

NEW YORK • LONDON

To those who still have the memory of

wonderful peaches, sweet raisins, and family farms.

I hope your stories are shared before they are lost.

Copyright © 2003 by David Mas Masumoto

Printed in the United States of America
First published as a Norton paperback 2004

For information about permission to reproduce selections from this book,
write to Permissions, W. W. Norton & Company, Inc., 500 Fifth Avenue,
New York, NY 10110

Manufacturing by Quebecor Fairfield
Book design by Dana Sloan
Production manager: Julia Druskin

Library of Congress Cataloging-in-Publication Data

Masumoto, David Mas.
　Four seasons in five senses : things worth savoring / David Mas Masumoto.
　　p. cm.
　ISBN 0-393-01960-8 (hardcover)
1. Masumoto, David Mas. 2. Peach growers—California—Del Rey—
Biography. 3. Farmers—California—Del Rey—Biography. 4. Japanese
American farmers—California—Del Rey—Biography. 5. Farm life—
California—Del Rey. 6. Family farms—California—Del Rey. I. Title.
　SB63.M36 A3 2003
　634'.2584'092—dc21

　　　　　　　　　　　　　　　　　　　　　　　　　2002010216

ISBN 0-393-32536-9 (pbk.)

W. W. Norton & Company, Inc., 500 Fifth Avenue, New York, N.Y. 10110
www.wwnorton.com

W. W. Norton & Company, Ltd., Castle House, 75/76 Wells Street,
London WIT 3QT

1 2 3 4 5 6 7 8 9 0

Contents

☙

v

The Journey
of a Peach

Things I Remember

✧

I remember peaches with aroma like a natural perfume, juices that dripped down my cheeks and a flavor that was slowly savored.

Nectarines with a tanginess dancing on my taste buds.

The lingering scent as grapes dry into raisins all through the month of September.

Picking vegetables with my grandmother in silence because she spoke no English nor I Japanese. Her muscles, smooth and round as eggs, had grown strong over the years.

The smell of my father's sweat.

Falling asleep driving a tractor and crashing into two grapevines. Now and forever, two younger vines stand next to old ones.

When the start of school in September was delayed to make time for the raisin harvest.

Holding a burning newspaper under a grape vine canopy to smoke out feisty wasps so we could pick the grapes.

Laying my sleeping children under a vine while I pruned and sitting them upright so they could watch me tie and rope weak peach limbs.

My normally quiet father once being so furious with low fruit prices that he chased a salesman off our farm.

Buying a tool and expecting it to last a lifetime; fixing old equipment instead of throwing it away.

Our family leaving handprints and our names in wet cement. I can still locate the names in the old barn's concrete floor from the farmer and his family who lived and worked here before us.

Kitchens filled with aroma that drifted out the window.

Farmhouse parlors.

Giving directions to strangers using landmarks like a farmer's stack of thousands of wooden raisin trays or his heaped junk pile.

A twenty-pound box of peaches selling for two dollars in 1961 and for two dollars in 2001.

When pesticides were so innovative it took years to convince farmers to try them.

The smell of manure spread in the fields after a good year, a farmer returning profits to the earth.

Families who canned fruit. My wife still makes jam every year.

When everyone wore an older brother's or sister's hand-me-downs as a kid and thinking my family was rich because I never had to wear my big sister's shoes.

Farmers who knew the same neighbors for their entire lives.

I remember when generations knew what good peaches tasted like.

Homeless
Peaches

Does Fast Food come from fast farming?

Ingredients for a peach: Well-drained soils. Balance of trace minerals. A few spring showers. Dreadful summer heat. Good cold winters. Organic matter. Indigenous grasses (weeds). Legume cover crops. A farmer father. Family generations working the land. And luck.

Peaches are supposed to be grown slowly, and I don't try to change that rhythm with large dosages of synthetic fertilizers or chemicals designed to stimulate growth. I believe my California soil adds a distinctive local flavor. My dirt — a sandy loam — is a peach's delight. Mixed with water from the Sierra snow, the blend is rich but light, not like the dense texture of clay. The weather patterns of the San Joaquin Valley are perfectly suited for my Sun Crest variety, which ripens around the Fourth of July, a good peach time in the middle of our hot summers (peaches originated in the deserts of China — they like dry heat).

However, efficiency, productivity, and speed have become the symbols of success in farming. As a farmer, I face a challenge common to most industries — a growing division between big and small.

Agriculture has become more of a business confined by narrow definitions of success and price-as-bottom-line thinking. With the industrialization of farming, farmers manufacture products, and decisions are made more and more by managers and not by owner-operators. Neighbors talk less and less with each other; we're too busy sprinting from one chore to another. We've lost the art of roadside talks – "collaborative practices" as an anthropologist friend once called it – and the gatherings of pickups and farmers sharing stories.

I rarely hear the phrase "farming as a way of life" to rationalize our livelihood. Instead I sense that our farms are becoming impersonal, that decisions are now based on business plans while personal responsibility is exiled from the fields. Industrial peaches are not grown but made with minimal variation. They've become a commodity produced for mass consumption.

One summer I tried to follow my peaches to market. We farmers worked so hard and for so long to grow, nurture, and ripen these precious gems, only to see them loaded on a truck and shipped off to parts unknown. The journey of my peaches lacked a finale, a sense of accomplishment. This didn't seem right, so I planned to monitor my peaches from farm to store and, I imagined, eventually a human hand and salivating mouth. Watching someone bite into one of my peaches, with a look of rapture and surprise, the nectar exploding and drenching taste buds, could make all my work worthwhile. I needed to know how it happened.

But I lost my peaches just outside of Bakersfield, California, about one hundred miles south of my farm in Fresno. Actually, the peaches continued on their journey, but a climate-recording device inside the truck container – a machine that's supposed to document the temperatures at which the fruit are maintained during shipment – broke. So my fruits momentarily vanished with no clear paper trail. I'm sure they arrived fine, since I didn't hear any complaints, but my effort to follow them failed miserably.

I lost them because they had become industrial peaches — a load of sixteen hundred boxes, each weighing twenty-three pounds, with a lot number and USDA inspection stamp and a faxed delivery tag with directions to a grocery warehouse where they'd be received, distributed to regional stores, and finally displayed. But I believe my fruits don't fit the industrial marketplace. Instead I prefer to think of them as "microbrewed peaches."

My strategy follows the similar phenomenon of microbrewed beers as they defied the established order in the world of beer, which had long been dominated by a few industrial brands. Specialty brews are made slowly and carry regional identities. They have original flavors that personalize their character. Many have a loyal following who thirst for real taste and challenge assembly-line "Spam" beers. A mutual passion drives both the brewer and the consumer.

My peaches embody the uniquely brewed taste of a small family farm. I can't grow them in large volume, nor do I want to. I work side by side with family members. I grow my fruits organically, and the process is intended to be complex because I work with an unpredictable mother nature.

No wonder I lost peaches in Bakersfield (if not there, I'm sure someplace else), because they were no longer Masumoto peaches. They had become a number and order, reduced to simply a "thing." Once my peaches entered the conventional food distribution system, I became a supplier of raw materials destined for markets that I need not bother to know. My farm was a factory in the fields, and my job was to forge something that looked like a peach, a bright red peach if at all possible, pack them in boxes, and kiss them goodbye.

Many years I've struggled just to bring a crop to harvest. One year it seems a worm pest rears its nasty little head and gouges deeply into the flesh. Another year an invisible pathogen makes itself at home in my fields, stealthily waiting until the right amount of heat during the day and sugar in the fruit, manifesting just when my hope for an ideal harvest begins to soar. Weeds pompously grow before my eyes, laughing at my feeble attempts to shovel them. Or workers don't show up and those that do seem to labor in slow motion.

During the heat of summer and the headache of harvest, the euphoria of that early-spring morning when the dew sparkled on the fresh leaves and the crop seemed pest-free and perfect seems like a distant and mean-spirited illusion. At times, I just want to get this over, especially if prices are lousy. Pick and push fruit off the farm. Hang my head and wonder what I did to deserve such suffering.

I imagine sometimes my wife, Marcy, thinks it's okay to lose my peaches, especially around July 2. My son, Korio, happens to be cursed with being born on that day in the middle of harvest, when I sweat daily over softening peaches and tractors stuck in the fields with dead batteries, or a box company that won't be working on a holiday even though peaches don't observe the Fourth; or my initial terror when I discover a single worm found on a branch, then anguish wondering if that was the only one when I can't find any more. When Kori was a toddler, I could fool him by claiming his birthday only fell on a Sunday afternoon and a rare day off during July. As he matured, I could still bribe him with an early celebration in June and a favorite toy.

His wonderful understanding nature makes me cringe, though, when I ask if we can "do his birthday some other time because of the peaches" and he gently answers, "That's okay, Dad." Marcy will follow with an icy glare directed at me. Now I feel guilty.

Hand-to-Hand
Journey

T HE NEXT SEASON, I tried to follow my peaches to market as they traveled from field to consumer. I imagined they journeyed from hand to hand, gaining in value en route, each step a link in our food chain.

My peaches begin with family hands. I suppose I can retrace the lineage of our peaches to a time when Dad first planted these trees three decades ago. Then, to truly honor the past, I should thank all of the prior farmers who have worked this land and contributed to the soil with their hand labor. And what about all the plants and creatures who gave their life for such a rich organic soil? But I'm not that good a Buddhist nor historian, so I tend to think of the immediate.

Throughout the spring, from thinning to the first blush of color at harvest, our hands probe and massage these peaches. We don't leave many fingerprints. Most of the time we're cradling the gems, feeling with our palms, holding them up to sunlight, watching and monitoring. I think of my peaches as handmade – personalized with my family's touch, a touch I hope translates into taste.

The journey of my peaches begins with a hand-crafted flavor, born from a regional advantage: the common geography I share with the other family farmers, a proximity that promotes interaction and exchange of information. We hope to learn from and trust each other. Living in this community can become a competitive advantage. Just as my peaches can't be grown elsewhere, farmer-to-farmer friendships can't be replicated and our farm stories can't be duplicated. Small, specialty farming depends on collective learning within a competitive structure. A chance meeting in town, at a dinner or funeral, provides opportunity to pass on information about market demand, coming legislation, labor trends, and regulations. We exchange our own "consumer's reports" on equipment and new innovations, especially with pest control. Solutions from the established industry need to be questioned; many of us have learned firsthand about the limits of conventional scientific research.

FARM HANDS

Hard physical work is necessary to grow a peach, then move it from a tree and field to a box and truck. The unseen hands of workers prune trees, thin and weed as the fruit grows fat, then come to pick, sort, pack, and stack. During one season, a million pieces of fruit from my small farm are nurtured, examined, and fondled by these hands.

Experience matters; the history of those who work the land provide a wealth of knowledge that continues to multiply. Each season, we seem to experience some sort of unpredictable weather I have never experienced before. The workers remind me of a winter freeze from the past or the time they harvested peaches in the rain. Once, during a ninety-degree day with humidity so thick that sweaty T-shirts and work shirts would not dry, a worker described how peaches ripen the best in such conditions. "Good for peaches, bad for workers," he said, wagging an index finger at me. He was right, and retailers told me so.

Ideally, sweat equity is valued and passed on as a peach journeys to market. A problem remains, though: while I do value the hand labor of farm workers, I don't set the price for peaches. Instead, I'm part of a system of sales based on supply and demand, and I often conduct a type of reverse auction – when there's lots of fruit around (which happens most of the time), instead of selling to the highest bidder, I hand over my produce to the lowest bidder who will take the most fruit. I have been informed this is euphemistically called "good movement." Lost in this journey is the craft of the individual laborers – their contribution to taste quickly becomes invisible.

HAND SALES

Hand stretched outward, palm up, ready to take my money. This is the image in my mind when farmers start talking of "the middleman," those who may never physically hold a peach and yet send tens of thousands of boxes along their way. But my perception is simplistic and a bit naive. Most farmers don't often sell directly to consumers. (Farmers' market folks and those participating in community-support agriculture are the rare and wonderful exceptions.) Middlemen and -women are a necessary part of the food chain. They have the contacts, can move product, and from what I've learned, tell very funny jokes – a crucial art when negotiating prices.

The jokes remain crude, archaic, politically incorrect, and very human. With prices shifting by the hour, inventories changing with a weather front, and anxious farmers depending on a price swing of a few cents per pound, the best brokers represent who and what I am. The humor helps everyone cope with the tension and stress of the moving targets. As Steve, my broker at Pacific Organic Produce, describes: "We're dealing with a time bomb because it's a perishable product. Everyone needs an emotional release and jokes help keep the edge off of it all."

The art of negotiating a fair price with individuals with whom you do business season after season must be a sticky challenge. The community of produce people, sellers and buyers, is quite small, often with the same faces moving within the industry. A seller one year may be a buyer the next, a contact becomes a competitor, a competitor a buyer. "We inbreed," said one broker. "And have long memories."

From a farmer's point of view, in a good year (read: low supply and high prices) the brokers' 10 percent commission seems awfully high. Yet trying to find a home for a peach in a bad market year or a season with pest problems makes a broker's efforts invaluable. "Sell it or smell it" becomes the motto as players turn frantic and desperate and fruit is drastically discounted before it starts to decay and rot and then must be dumped out into our fields. The feeling sickens a farmer; we're so poor we have to bury our own dead. Give me crude jokes, solid sales, and "movement."

I've been fortunate to find brokers who care and understand the biography of my peaches; after years of auditions, I found a team who want to be part of the story. Most brokers do fine when prices are high, but great brokers perform in tough years, negotiating hard when hard matters yet sharing information when I need to hear bad news.

"Customizing," one of my brokers calls it, working to ensure a proper match between my produce and a buyer. The term also refers to my practice of "just in time" management (which conveniently serves as a rationalization when I'm behind in work). Schedules are in flux and remain necessarily flexible as we adapt to changing conditions.

I sprint to adjust while trying to understand the slowness inherent in my work. I respond to the daily whims of weather or the seasonal shifts that delay ripening or speed up harvest projections. Long-term calendars simultaneously stand watch with my hourly work pace. When I planted my orchard of early-season peaches, the initial three years to first harvest seemed like an eternity. But the following summers of inconsistent crops led me to conclude that even though the orchard was filled with ripe peaches, they behaved like adolescents, unpredictable and uncontrolled. On the other hand, my Sun

Crest peaches have been with my family for over thirty years and I've gleaned enough experiences from their many seasons to guide my decisions. A great marketing team knows all this and handles my confessions of turbulent workdays along with pallets and boxes.

The last time I see my peaches is when a forklift carries a pallet through a thick plastic curtain and into the dark caverns of a cold-storage room, temperature a chilly thirty-five to forty degrees. It's the end of a July day, I'm beaten and tired. I chant to the gods who oversee my brokers, shippers, distributors, and retailers, "Do your magic" – as I limp home.

Then in order to follow a peach to market, perhaps the simplest method is to follow the money. Consider the classic pie chart, with slices representing the different hands that pass a peach along. Assume the whole pie is worth a dollar, a dollar a pound for peaches (already some will claim that's too expensive). Farmers are paid last, so let's skip their slice but return to the other sets of hands on the farm and their slice.

Farm workers contribute vast amounts of labor in growing and harvesting a peach. Their work is stretched out over nine months, beginning with pruning in the winter, thinning in spring, and then shoveling, irrigating, and weeding as required – all for eight cents per pound. At harvest, they walk an orchard numerous times, climbing ladders, searching for and hand-picking the best, and returning a few days later to repeat the process – add six cents to the total. Sorting and packing fruit, palletizing the boxes, and strapping and delivering them to the cold-storage facility can easily run ten cents a pound. Just for hired field labor, a peach can cost twenty-four cents a pound. Our operations remain extremely labor-intensive, little factories with dozens of pairs of hands creating value, something with flavor.

I try to keep my costs to grow a peach to a minimum, usually substituting management for capital. I've been fortunate in avoiding major mistakes and reducing where I can. Pest control materials, organic fertilizers, compost, and cover crops add up to a rather thin slice of the pie, only about four cents per pound. The box in which we ship the fruit, along with packaging materials, can run about ten cents – an expense that can't be avoided. Fees for other hands – for USDA inspection and industry promotion program (we tax ourselves about a penny for every pound of peaches); cold storage, three cents; and commission for the brokers, ten cents.

So to grow and harvest a pound of my peaches, to move them from field to cold storage and have them ready for shipment, will cost about fifty-two cents out of a dollar. Machines can't duplicate this journey. Each set of hands adds a service, a value measured by money. Half a peach's worth is already sunk in production and harvest costs, with nothing yet in it for my labor. Meanwhile, the boxes still sit just a few miles from my farm.

A HAND IN HAULING

A little over a century ago, Brooklyn was the garden for New York City, supplying vast amounts of produce to feed a thriving population. According to Marc Linder and Lawrence Zacharias in *Of Cabbages and Kings County*, the fields of Brooklyn withered with new refrigeration techniques that made it cheaper to ship vegetables from the Deep South. The shippers – trains at the turn of the century and trucks today – changed the face of agriculture. They were revolutionaries.

It's hard to imagine truck drivers as radicals. Some will say the modern transportation industry has not resulted in progress but rather creating an unnatural consumer taste – the ability for someone in chilly Boston to get my peaches in May, only possible because of these rebellious truckers. Perhaps. But the opportunity to obtain quality produce in a cold climate can outweigh the old "root cellar" method of aged and slightly moldy potatoes and apples stored over a winter that Marcy's Grandma Rose in Wisconsin used to describe to me.

Insurgent truckers pick up my peaches and head for a journey over our highways. West Coast deliveries in a day or two, Midwest in three, East Coast a four- or five-day haul with little sleep. The cost to ship a twenty-three-pound box of peaches ranges from $2.50 to $4, depending on availability of trucks, pickup and drop locations, and destination. Add another cost to our peach pie — a ten-to-fifteen-cent slice.

While this sounds outrageous — especially considering the farmer who grew the stuff for months and months probably won't get that much — the process is much more complex than at first glance. First, most produce trucks have to travel to multiple places to pick up produce, gathering some peaches from one cold storage facility, nectarines or plums from another. Also refrigeration equipment must operate correctly and can never be shut off during the multiple-day run from one coast to the other. Finally, truckers haul a perishable product that must be delivered within a window — often a California order is loaded on a Monday and expected to arrive on the East Coast by Friday morning in time for distribution to weekend shoppers. Miss a drop time and orders are lost. If you calculate the required rest/sleep time for the driver, there's almost no room for error, unless a few log entries are altered, which I'm sure happens. Add in highway taxes, overhead, equipment expenses, and the largest single cost, the diesel fuel, and no one gets rich doing this. Truckers are another set of hands, literally hauling the fruit to market.

DISTRIBUTORS' HANDS

Some peaches journey directly to the retailer, usually a chain store, who can buy a truckload (about fourteen hundred twenty-three-pound boxes) at a time and ship to their own warehouses, where the fruit is allocated to numerous stores. By skipping hands, some expenses are reduced, yet fruit still must be received, stored, cooled, handled, then dispensed. The majority of my produce passes through an often invisible set of hands, the distributors. They act as a go-between, a type of produce matchmaker linking luscious fruits with perspective suitors.

Distributors, sometimes called wholesalers, are expert puzzle solvers. They take pride in sourcing hard-to-fit pieces – a small natural foods store, an exclusive restaurant featuring a menu with organically grown fruits and vegetables, or a specialty produce grocer. They work on a micro level – I've seen a buyer pull out from their facility in a station wagon, backseats loaded with a week's supply of peaches, certainly not what you'd call volume sales.

When my peaches first arrive at a distributor, they enter a staging room, where counts are verified, quality is examined (if there's a problem, after a series of phone calls, adjustments in price are made), and product is "slotted" or inventoried and stored in a small warehouse, typically on-site. The storage facility can be very complex. Roots and Fruits, a distributor in Minneapolis, has eight different rooms for eight different temperatures, each for a specific category of produce. Stone fruit, peaches, plums, and nectarines, need a temperature near thirty-two degrees, much too cold for leafy vegetables like lettuce, which could easily freeze in such an environment. Bananas have a warmer space often called a ripening room, and tomatoes join them there. Apples and carrots are stored in another room. The process is elaborate and complex, well suited for the customer service they provide.

I know the least about this set of hands. Distributors work behind the scenes, almost in a mysterious manner, perhaps because at this point in the journey, I can claim my peaches in name only. The wholesalers own the fruit, but both of us benefit from quality and wonderful taste. A respected name and reputation at this milestone isn't just a feel-good expression – it translates directly into sales and income.

The distributor's share? It can be high – as much as a 40 percent markup. I'd safely add twenty cents to my peach pie chart.

IN THE HANDS OF RETAIL

Retailer – the final set of hands before a peach reaches the customer. The produce section of a grocery store has evolved as our tastes changed to a healthier diet of more fruits and vegetables – track lighting for visual accent, sophisticated misters and washers to keep produce fresh, multiple restocking and maintenance to ensure a groomed appearance. Produce is often the first department a shopper enters, appealing with colorful and artistic displays that invite and attract. My peaches find a home here, alongside other tantalizing produce.

Produce requires many hands at the retail level. Hands unload a truck, sort and store goods in a walk-in cold storage. Other hands stock and restock shelves, often numerous times daily, cleaning up stands, adding fresh product, removing the old. Produce has a high shrinkage factor, because the product can quickly grow old and some is damaged by shoppers. All this is tolerated because produce has grown into a major profit center, brings in consumers, and grocery margins, typically about 2 percent, are often higher with fruits and vegetables. This poses a problem (from a farmer's point of view) in today's competitive climate. Sometimes the power of produce to pull in extra profit – I've heard it can be 5 percent or higher – is used to support other less profitable sections of a store.

But every harvest, there's a time in the season when the field price tumbles for peaches, plums, or nectarines. I will be forced to let go of my peaches for desperately low prices. Assuming other costs are the same for trucking and distribution, I can estimate a retailer is getting them for as low as forty cents a pound. I'll then wait for a grocer to "go on ad" and purchase advertising space in a local newspaper, helping the industry with the glut of fruit, moving product that's already deeply discounted and still making good margins. Yet many times prices stay the same, a dollar a pound or more. Sales remain flat, a retailer's profits soar, and farmers fume.

Then along country roads, farmers will stop to gripe and complain. At the packing shed, they'll cross their arms and shake their heads while their trucks are unloaded. We can do the math. We're hurting. Fruit will be left on the trees. All the while, some retailers hold their prices at two or three or four times their expenses. Peaches become a cash cow – for someone else. And there's nothing I can do about it. The produce aisle – the one section of a grocery store where the human senses of smell, touch, sight, and in some cases sound (thumping a melon to judge ripeness) come alive, the arena where high touch and sensory perceptions count (ever fondle a box of cereal or smell the plastic bottle of soda?), this section filled with my personalized produce – subsidizes and finances dozens of other aisles.

I suppose there are times when the economics of a produce section don't meet expectations and the rest of a store helps maintain our presence. But two observations compel me to think otherwise. First, produce is one of the most fluid areas of product lines. Peaches are quickly replaced one week by grapes, followed by bananas or apples and pears or something else another week. A poor-selling product can be relegated overnight to a small, obscure corner, and shoppers are used to such changes. Move peaches off the choice display at the end of a table or stand and no one complains. Move the diet sodas from one shelf to another just below it and store managers are chastised for making the product too hard to find.

Second, business practices of other products have now found their way into produce. For example, in the dry goods sections of a store, the controversial policy of slotting fees has allowed stores to charge for shelf space: prime locations at eye level are rented to various companies, and the suppliers pay for their space. Grocers gain revenue as competitors vie for visibility. Major corporations fill these slots and in some cases add a product in order to ensure visual monopoly of a shelf. I have always wondered why there are so many different types of Jell-O and question if people really do buy that much watermelon flavor.

The produce section, and specifically stone fruits, has been resistant to this practice, perhaps because of its already healthy profit margins and also the lack of a dominant player in the industry. In the past, there has not been the one or two major corporations who may bid for visibility in order to drive the others out of the market. But that is changing as more and more stores charge companies for display space by the "slot." Retailers see slotting fees as another revenue source. The system of supply and demand that my prices are based upon seems old-fashioned in this new world of slotting fees, market visibility, and profit centers. Slotting adds a new set of hands my peaches must pass through, hands by whose neglect my fruits might quickly be lost.

This results in a bigger piece of the pie for retail. I'll guess about ten cents a slice, which sounds low at first, although the retailer is the only one who has the power to set a final price. While it can be said that a customer ultimately determines what is bought, grocers can always substitute lower-priced mangos or other fruit for peaches to protect their bottom lines. This helps ensure that their investment will usually be covered and result in a profit. A grocer's hands are very versatile; mine are bound to the tree stuck in the ground.

FARMERS' HANDS

Sending my peaches to market means letting them enter a world filled with verbal handshakes and promises as fruit is handed off or lateraled to another pair of hands and another. Each set of open palms promises a favor for the farmer, by taking the crop off our hands — for a slight fee. The touch of family hands is the only one that doesn't set a price. Instead of costs passed on as a product works its way through the maze, I begin with a whole pie with pieces gradually cut away, and if I'm fortunate, I'm left with a final slice, a very, very thin slice, six cents a pound, according to my calculations.

Suddenly a penny or two at the retail level makes a big difference – assuming it's passed along. Farming by pennies is a hell of a way to make a living. What happens in those seasons when prices collapse and I lose money?

Costs incurred along a market journey can't always be tacked onto our final price, and the service each set of hands adds doesn't necessarily add to the monetary value I receive in the end. In bad market years, players want to wash their hands of these peaches. Debt and negative cash flow – is it possible to have a negative peach?

I know for sure that a peach that doesn't earn money can still have taste. That hasn't changed, only how their value is measured did. I'm not sure my peaches belong in this hand-to-hand economy. I hope consumers appreciate their value – a value beyond prices.

The Story
Economy

I LEARN HOW TO FARM from stories. Uncles, aunts, and neighbors offer opinions and complaints – we share our successes and failures. Spontaneous discussions, some brief, others prolonged, over food at a family dinner or leaning against a truck radiator on chilly mornings when delivering raisins. Publicly, many farmers appear stoic and reserved, our speech relaxed. But we know much is conveyed through the unspoken: a pawing of the ground with a work boot, a slap on the back, the slow shaking of a head back and forth, followed by a piercing silence. We may seem dull at times, yet as we exchange tales, we are teaching and learning from each other. Daily "class" is held at coffee shops and through roadside conversations. It takes years, perhaps even decades and generations, to learn how to grow a wonderful-tasting piece of fruit on a specific plot of land, to match the right variety of peach to the soil texture and type. The face-to-face exchanges remain invaluable; opinionated neighbors become assets. Knowledge of my father working the land becomes the greatest resource I have. I welcome anecdotal information. Though our produce may be bound for global markets, the solutions to our farming challenges remain local.

I manage by stories. My fruits begin their journey with a history of tales, a series of linking poems about the earth and human hands.

MEMORY ECONOMY

During an in-store tasting in San Francisco, I offer a slice of one of my peaches to a twenty-something-year-old, and the young professional (imagine a computer whiz, alternative lifestyle, flushed with wealth) breaks into a grin, eyebrows raised in disbelief, with a look of surprise, no, shock, over his face. He has never tasted something like this before. I am quite satisfied to have created a new memory. Next, an elderly couple enter and taste. A grin, eyebrows raised but not in disbelief; something deep is stirred. Great peaches elicit almost forgotten sensations, evoking a smile of recognition, not skepticism. Stories flash through their minds, images of a backyard tree, a grandfather's farm on summer Sundays, a family gathering. My peach travels into their past and brings on a journey of transformation, even if just for a moment.

I claim that my peaches fill the flavor niche industry left behind. Large-scale farming operations can't mimic my methods, in which skill and human management replace huge doses of capital and technology. I want my fruits to manifest the life and spirit of our farm. Mass-produced peaches are designed to excite the visual sense as consumers trade money for something that resembles a peach. But my peaches begin with a journey into taste, texture, and aroma, accompanied by stories. People who enjoy my peaches understand and appreciate flavor.

How do I know my peaches taste good? That's what my quest to follow my peaches to market is all about: feedback. I want to hear the rest of the story, to learn of the routes my fruit ventures to market, the hands they pass through, and even the response from the ultimate end . . . eater? I don't feel comfortable with the term "end user," nor do I like "consumer"; both sound terribly impersonal. I like the term "audience" for my peaches, something alive, with feelings on everyone's behalf. I'm more of a director, nudging character from my fields, orchestrating emotions and action. It's a business and art.

The audience for my peaches pay attention to their foods. They care about meaning and, given a choice, are interested in difference. I offer options, a dialogue between producer and consumer. I hope to create a new appreciation when my audience can take it – and it means a story – with them.

I grow the way an actor performs – looking for an emotional response. My fruits are not simply a good or service. Emotions matter, and the story of how I farm and the journey of our foods to market is the central plot of the performance. But I am not the kind of performance artist who creates and expresses, then leaves it up to others to discover meaning: "I grow it. You eat it. Be happy." I think of each year on my farm like a novel, a story unfolding, shared and reviewed – a peach as a communal reflection, not just individualistic expression.

When I follow my peaches to market, perhaps my goal is not to demystify the process but rather to make it more complex by adding a highly subjective quality of experience to the journey. My story now accompanies each piece of fruit.

The more others know of my farm, the greater potential I have for success, especially on a personal level. When a farmer's story goes public, it can affirm farmers' work because outsiders grant new recognition, a type of legitimacy. Easier ways to make a living exist, but knowing my produce and sharing that knowledge with others satisfy a hunger and reward me: my family's work has meaning.

A sense of place — where food comes from and who grows it — becomes an asset, not just a promotional tool. Sometimes I'll see ads for mass-produced commercial foods with the image of a lone farmer walking fields of grain, sunset creating silhouettes of gold, then cut to a healthy family smiling with a voice-over, "From our family to yours . . ." The character may or may not have been a farmer, the fields may or may not have been actually his or her own, and the emotions may seem real for only a second. But that's usually it — just a sensory flash-bulb and then it's gone and we're left blinking. I don't think many consumers think of that farmer when squeezing a loaf of bread. That slick ad lacks an honest flavor — a bittersweet taste left in the mouth because we know the real story.

The best farmers of personalized produce strive to create true stories and personal connections through our fruits. While we may occupy only a very, very small slice of the industry total, we don't want to change the whole world, but we do want to make a little room on the shelf for our flavors. Every year I still try to follow my peaches to market and follow a story of taste — to end with the slow trip from hand to mouth and palate to memory. A journey through four seasons in five senses.

PART ONE

The Art
of Seeing

First Steps into an Orchard

Old orchard. Dark forest. Dense green. Layers of leaves. Shade. Coolness. Calm. Brown limbs – coarse. Tiny fruits – delicate. Shoe prints left behind. Soft earth. Clovers. Wildflowers. Natural grasses. Silence. Buzz. Insects. Life underfoot. The underside of a blade of grass. Weeds sparkle with dew. Wild Italian rye shimmers. Sun strikes. Rays spiral through the canopy. Powder-blue sky. Fresh buds. Pale green. Thickening shoots. Fat peaches. Soon.

Planting
Trees

MY OLD PEACH orchard tells our family's story. In the twisted trunks lie the history of my father, who planted these trees more than thirty years ago. I recall helping him as the family lined up the trees by sight – holding upright a bare-root tree, closing one eye and squinting with the other, peering down a quarter-mile row, eyeballing wooden stick markers spaced throughout the orchard, and trying to keep the rows straight. We weren't perfect nor fast, and for decades I've had to turn my tractor and swing wide in one direction to avoid striking a crooked tree. But planting five hundred trees by hand and trusting our vision seems to be a wonderfully human way to begin an orchard. We made mistakes and rationalized our efforts – life in nature is not always straight.

Most farms today keep peach trees only about ten years before replacing them with a newer variety that produces bigger and redder fruit and more often. But I consider my aging block of trees – a wonderful-tasting variety called Sun Crest – my "old growth" forest. In the base of some trunks flickers have pecked a hole and carved out a nest. Amazingly the tree survives, enough of the cambium layer unaffected for the present by the bird's sharp beak. Silly to keep a damaged tree – production will be reduced, the life span probably cut shorter. Yet I find comfort in knowing the creature nests comfortably, its home safely tucked inside one of my trees.

By no means are the trees the oldest creatures on our farm. We also work sixty acres of raisin grapes with thousands of vines more than ninety years old. They still produce well despite their age. Over the years I have concluded they too were planted by sight. Many rows are crooked, and others that begin wide, with twelve-foot spacing, narrow to about eleven feet a hundred yards down the row. Every few years I replace a few vines, some fallen with age or damaged when farm equipment has accidentally hooked a gnarled, misshapen trunk. I'm tempted to replant them all with the proper spacing, but I don't. Instead I slowly add to the crooked timeline with my replacement vines.

As a child, I watched Mom and Dad trudge home from the fields daily, sometimes happily with the scent of harvest in the air. But other times they returned late, long after the sun set, physically exhausted from the seemingly endless work. They never talked much about their work. Instead I witnessed the long hours of physical labor. I have memories of watching a flickering set of tractor lights traveling up and down row after row in the darkness, bouncing over uneven fields, as my father raced to finish disking before rains arrived.

I worried about the physical work gradually stealing years from my parents. I recall the daily hardness of the farm, the impoverished struggle. We couldn't afford better machines to ease the pain; we saved money by doing the work ourselves. Mom never learned to cook or sew well, but she could handle a shovel or pruning shears as a veteran. Dad had little time to play. There was no romance of a father and son playing catch; we had no annual rite of attending a baseball game. Life was slowly sucked from our family, so gradually I'm not sure we realized it or even wanted to know.

I remember one year's first heat wave in May. Mom's face became flushed red. She made a farm woman's white bonnet from used rice sacks to protect her from the sun. It helped little. I remember how Dad's face turned darker in June, especially the year we lost half our crop when, with intense temperatures over one hundred degrees, the trees aborted their fruit. As the thermometer soared, he had no quick fix. He rarely spoke with emotion, but I saw the pain on his face when he didn't want to go back into the orchard to assess the damage.

I knew my parents' work. They made no attempt to protect us children from the harsh realities of farming and its tremendous risk. I learned of the uncontrolled and precocious spirit of nature that treats us one year with juicy peaches, then robs us the next with an outbreak of brown rot, only to have the cycle repeat, teasing us with a good year followed by a series of disasters.

I grew angry. We had little control over our lives. We labored long hours for meager profits. No one seemed to give a damn for our fruits. Everyone chased inexpensive food. I deemed our worth had little value and the rhythms of our work out of sync.

Roping
Trees

Peach trees take time to get to know. It has taken decades and
thousands of mistakes to learn how to prune these trees. I don't mind
my errors — they help me understand the art of shaping a tree.
Throughout the course of a year, I'll visit many trees over and over,
clipping and bending branches like a *bonsai* artist. I've learned that
young limbs will push hardy shoots and strong wood. I prune those
"hard," a process of selecting the best and culling out the weaker
wood, trusting my judgment that by next spring the remaining
branches will be lined with healthy blossoms and eventually strings of
luscious fruit.

With the older limbs, though, I sense a type of maturity. Their
shoots are shorter but more numerous; they grow gradually, with mod-
eration. So I leave more wood, knowing each of the short stems can
handle only one or two peaches. I adjust to each branch, adapting my
method and approach to the character of the tree. On a few trees I tie
a small ribbon, a reminder to monitor the growth and defer my decision to prune hard to another time, later in the season. I plan to return.

Yet I have much to learn. A few years ago, I planted a new orchard of young peach trees next to Dad's house. After their third leaf, the trees stood at eye level, the growth vigorous.

"Wild," Dad called them.

"Young," I responded.

Dad spent a winter in that orchard, pausing at each tree. His old eyes darted from limb to limb. He pruned, snipped, paused and tilted his head like an artist trying to gain perspective. Then he stepped back, circling the trunk, clipping only where needed, engaged in a dance while shaping nature.

Once he stopped and searched the ground for a thick piece of wood. He snipped off all the lateral growth until he had a two-foot-long stick. He turned to a young, odd-shaped tree and pulled an awkwardly positioned branch toward him; he then wedged the dowel between two other limbs, forcing the growth outward.

Another time, Dad pulled off his gloves and grabbed hold of one precariously angled branch. Gently he bowed it back, shaping in his mind's eye a tree with a center that opens into a vase and gently curves outward, creating strong scaffolds for future pruners. Satisfied, he let the branch go, mentally noting the position that could be fixed in years to come; with other trees and an angle too great to manipulate, he would clip the branch out.

I finally noticed his secret technique. Tree by tree, he completed his work by lashing a series of ropes from branch to branch, crossing the center, pulling a vagrant back into place and balancing it with the rest of the tree. Dad understood geometry and had his own theory of triangulation for support. For some, he had to first tie two other scaffolds together, dissecting the middle, then knot a strand at a right angle, allowing the critical support to be fastened in a T shape. From a distance, his orchard looked spectacular, orderly with a natural symmetry. Close up, ingenious configurations of rope held the branches in check, adapting their positions.

As a child, I had first pruned with a pair of hand-me-down shears and I struggled to discover the "right" branches to leave. Later I learned how to make radical slashes with a chain saw and from these healthy cuts fresh wood grew. Now, after study under Dad, I add a new tool – ties that bind.

Dad pruned, shaped, and roped with a long memory, not merely for next year's crop but also for the year after that and the following year. He left a legacy behind, not in monetary fortunes or stories of success, but rather in old peach trees that have been cared for and properly shaped for a farmer's son. Dad could see the future. When I complimented him on his work, he answered, "This may be the last orchard I'll train." Then he smiled.

Blossoms

WHEN I WALK my orchards in early March, blossoms announce
the arrival of spring. The leaves, the true workhorses of trees and photosynthesis, remain hidden and quiet and won't fully appear for
another two weeks. They allow the delicate pink flowers to take center stage for the moment.

I decide to pause and study the bloom. My pickup truck rattles to
a stop in a cloud of dust; the engine sputters and protests, knocks and
shakes, then expires with a mechanical sigh. All is quiet. The lack of
movement catches me off guard; the sudden stillness makes me
uneasy. I know of weeds to shovel, plows to hook up, and tractors to
awaken from their winter rest. Instead I force myself to walk slow.

With the first warm temperatures in the seventies, blossom buds
grow fat and bulbous. A few have already bloomed. We have names for
the different stages of bloom. "Bud swell" comes first, then "pink bud,"
with just the tiny tip of the flower exposing itself, peeking through the
jacket. These dots of pink play tricks on my eyes — the orchards are
brushed with a subtle hue so delicate my eyes search for the source of
color. If I focus, all I see is the brown-and-tan bark. If I look away, then
look again with eyes wide open, the delicate pink veil returns.

Farmers mark calendars, but work commences only when nature

dictates. Even conventional, business-oriented farm managers pay attention to blossoms – for them, pink bud is a signal to spray fungicides for brown rot and blossom blight.

When blooms are just about ready to open, they're in the "popcorn" stage, a tight cluster of petals packed around a center ready to explode open. As the name implies, these blossoms are primed to burst with an energy and delight. Overnight they "pop," one kernel after another, a silent beauty even the most stoic farmer notices.

I think all farmers see hope in these glorious blossoms. Hope for the coming year. Hope for prosperity and profits. A day before, I watched my seventy-nine-year-old father blink and gaze over the orchard. I believe he saw hope because he's survived – through the chill of winter to the promising warmth of spring. He nudged me, pointed to the orchard, and gestured with his hand, holding a gap between his thumb and index finger, then flinging his hand open, mimicking something about to explode, something to happen soon in our fields.

I step into the orchard and guess we're about halfway through the bloom season. Estimating is an art, since the flowers blossom over a two-week period, depending on temperatures and sunlight. The blooms tend to start opening at the bottom of the trees and work their way up, the opposite of what I would have thought, since more sunlight reaches the top. But without leaves, the sun is democratic, and I imagine the sap and fluids of an awakening peach tree begin in the roots and head north, up the cambium layers. I walk and look up to where most of the blossoms dazzle. I try to quantify percentages of pink bud, popcorn, and open blossoms against the cloudless sky. I sometimes forget what I'm looking for. The thought of measuring the progress of bloom becomes lost in the cloud of perfectly shaped petals framed by the powder-blue canvas.

I refocus and whisper, "Fifty to seventy percent bloom," and jot down a note on my pocket notepad. In some years, precisely at this time, I'll apply a microorganism called *Bacillus thuringiensis*, a bacteria that emerging worms will consume and that will kill them. Different "Bt" treatments are formulated for specific pests and have a very short effective life – a day or two. I target the peach twig borer, which awakens with blossoms and begins foraging, hungry after a long winter's sleep. I'll try to apply Bt at 50 percent bloom and finally at full bloom, hoping to time my programs with the appetite of these larvae. New biological technology research has found a way to splice Bt organisms into the genes of certain plants like corn, producing a hybrid: pesticide in a plant. But I'm convinced this gene therapy defeats one of the benefits of organic farming: walking your fields and monitoring life. I want to be there when the pests break fast.

I've never counted the total number of blossoms on an average tree. Each peach tree stands ten to twelve feet high. Mature ones are loaded with slender branches trained to be exposed to the maximum amount of sunlight. I quickly estimate, three thousand, no, perhaps five thousand blossoms per tree and one hundred trees per acre. In my small five-acre block of Sun Crest peaches, at full bloom, two and a half million flowers? That can't be right – it sounds like too many. But staring down the rows of trees, I see layers upon layers of blossoms, a solid block of delicate pink.

Why are peach blossoms pink? Or plums white? Or nectarines – which are not a cross between plums and peaches – why are they pink too? Why not scarlet or yellow or orange? There must be a reason; nature doesn't work at random.

Or does it? White attracts bees better than pink, which is why plums and almonds, which need cross-pollinating, have white flowers, while most peaches, which are self-pollinating, are pink. Each of my varieties of peaches has a different shade of pink. The blooms of Elegant Lady, a peach ripening in late July, are dark and vivid, while Sun Crest flowers are light and soft, and Flavor Crest flowers are bright. But the colors don't correspond with the redness of the fruit at harvest. Furthermore, the varieties don't bloom in order of ripening. My earliest variety, Spring Lady, blooms just days before my latest fruit, harvested two months later. Sun Crest, a mid-season peach picked at the end of June and early July, is one of the last to flower.

I start to walk again, telling myself that it's okay to wander.

The intensity of colors varies from year to year. This year I've never seen a richer shade of red deep inside the bloom and surrounding the pistil, with a fresh pink running along the curves of the individual petals. The blossoms shine bright and vivid. Is this an indicator of a good year?

Anecdotal evidence tells me good harvests follow harsh winters. Mild winters may be good for humans, but peach trees wake up grumpy, tossing and turning, having never gotten a good sleep. Peaches long for hundreds of hours each winter with temperatures in the thirties; it need not freeze, but cold, damp, foggy days are perfect. Once we had a tropical winter when at Christmas the balmy weather allowed us to wear T-shirts and sweat as we opened presents. That season's peaches were grouchy – a bit mealy and lacking a sharp flavor. Even when ripe, they seemed to go flat and did not ship well.

In the early 1990s, just a month and a half before bloom, we had one of the coldest winter freezes I could remember. With the week of low temperatures, the trees slipped into a deep sleep and later, in early March, awoke refreshed. The blooms were strong and bright; a deep, dark red stain shone from an area around the pistil, a galaxy of pink with crimson flames at their centers. Months later, during harvest, we had some of the best peaches ever. They were sweet yet firm, stored well for a week and beyond, and had a deep blush when ripe.

I had forgotten about the blossom colors until I glanced at my notes at the end of harvest. Was there a relationship between color and quality? First I would need a list of terminology to use in describing the differences of blossom colors from one year to the next. Deep hues or subtle tints. Bold reds or hints of magenta. Intense pink that's vibrant one season, then soft and muted the next. Exciting and mysterious. Captivating and intriguing. Fun.

Peach blossoms do not have a strong aroma. Almonds do — I get headaches when they bloom, the fragrance so intense it must be impossible for bees to miss it — that is, if bees can smell. But the scent of peaches is subdued. I hold a bloom in my hand and breathe in the perfume deeply. I can barely detect it — alluringly subtle.

Walk slower now. Deep into the orchard, pink hues all around me. The delicate petals are translucent; they allow the sun's rays to pass through, diffusing the light, creating a soft luster in the orchard. Like stained glass — quiet and spiritual. I can feel a breeze whisper across my cheeks. A petal drops from the tree, dances downward, a cradle rocking, gliding to the earth. I can hear it land with a silence. The orchard feels like a cathedral, branches arched overhead, flower petals transforming light, fields stirring with life and spirit. For a moment I don't think about peaches. I see blossoms.

Thoughts of a dear friend, her spirit and smile as bright as a blossom, her will to help others as beautiful and strong as this old orchard. She has cancer. It drains her of life; she fights and may win some battles only to find the disease has extracted much. Test results, procedures, medication programs, setbacks. She lives from moment to moment, at times in pain and fear, with occasional pauses for small celebrations; bad days and, yes, some good days.

I think of her as I walk this orchard. Within weeks the bloom will be over and withered petals will tumble to the ground. Fragile and insignificant, they leave behind tiny fruit that will grow. This remains a sacred place where beauty lives and passes, a fragile moment that translates into an eventual hope of rich harvests. Blossoms glorious precisely because they are ephemeral. For the moment they are shining.

I begin to head back to my truck and other chores that I must rush to. This marks the beginning of the season, with much work to follow before harvest. Already a few petals are scattered on the ground. Each peach blossom has five petals, thousands of blossoms per tree, hundreds of trees in front of me — did I calculate millions and millions of petals for this one, single orchard?

Leaves have started to poke through their casings, showing the green that will soon dominate. With each breeze a shower of peach petals falls like rain. A layer will soon blanket the earth for a day. The orchard will again be alive in pink. Then a gust will blow and shift them. Some will be piled like drifts against the closest ridge in the dirt, others will wander to a neighbor's field, riding in the wind.

Snapshots

NEAR THE END of every harvest season I take some pictures. At times I feel like a wartime photojournalist, snapping pictures of fallen trees and dead limbs, broken by the weight of the fruit and old age. Covering the earth lie thousands of discarded peaches, tossed carelessly aside when a defect was observed. The carcasses float in my irrigation furrows, decaying into a mass of gray and hoary flesh. As I trudge down a row, I sense the battle is over, and in the next few months I will calculate the losses from the fields. But the fields are also filled with a lush density, a cornucopia of colors and life captured with each click of the camera. Some trees seem relaxed, relieved of the burden of their crop. The weeds sprout everywhere, nourished by a drink of water and decomposing fruit. Another harvest near complete.

I like to photograph the changes from year to year. I compare the bare, weedless earth of twenty years ago when I was younger and our farm was beginning to look sterile and lifeless to the present-day greens of clover and legume cover crops and the constant buzz and hum of insects working in an organic farm.

I wish my grandparents and parents had taken pictures of their work. I can only imagine how they fit into California's San Joaquin Valley landscape a hundred years ago, a quiet but hardworking farm couple from Japan immigrating to a new land with new crops. Later, did Dad and Mom work side by side with Jiichan and Baachan? When did the Masumotos first start growing peaches, and how often did we kids take our afternoon naps in the shade of a grapevine? I have no photographs of their work, only the special formal photos when family gathered for a wedding or a funeral, or an occasional community event like a picnic or community hall dedication. Cameras didn't belong in their fields.

I wander through the orchards, snapping images of a few trees still loaded with fruit and branches bending low, as if in pain, waiting to be freed of their harvest. The greens of the cover crops seem to become more vivid in photographs. I'm not sure why; perhaps when I walk through the fields, my eyes grow accustomed to the lushness all around me. The multiple layers of leaves and the dense growth of tens of thousands of clovers mixed with weeds blur together into a color I call "living green" – my own tropical rain forest or old-growth woods.

Voices of my workers carry through the orchard. They're picking the final trees in one corner of the field. Someone has brought a radio, and I can hear Mexican music dance through the greens. Another season comes to a close in these orchards where we worked so intimately for the last few weeks, picking round after round of fruit from each branch and tree. We will abandon these grounds and tomorrow move on to another crop and harvest. Yet for now, no one rushes to finish. I pocket my camera and return to my work, supervising the picking, hauling the harvested fruit from the fields to the packing shed and transporting empty wagons to the workers. On my return trip, I stop and take photos of the workers on ladders, the picking boxes and trailer full of peaches. The crew boss, the one picker who likes to have his picture taken, manages to be in a lot of the candid shots. I must amuse the work crew – they're all talking and joking to each other. My Spanish is so poor I can only catch a few words I know, *fotografia* and *bonito*, as they tease the photogenic one.

40

It occurs to me that I haven't asked permission to take their photos. I wonder if any believe that a photograph steals souls and imprisons them in the frames. No one has complained but I question whether they would protest.

I'm the farmer and these are my workers. Do they labor in fear of my ability to hire and fire? Or is it my responsibility to take care of them, a paternalistic relationship common to farmers and rural communities? I provide them jobs and wages; they do good work for me. I feel guilt at times, since they work harder than I. Same old story — those who do the real work are often not compensated enough. Yet most years, it's the same for the farmer. We work for pennies, and people in America spend a smaller percentage of their income on food than do people of any other nation. Cheap fast food mentality. Farm workers and farmers — we all become invisible.

The camera shutter clicks with the sun high overhead. Most of the workers' faces are dark, in the shadows. I want one of them to look up, move his face into the light, then I realize I've forgotten his name.

Photographs are often considered a luxury to these men. They work for low wages, migrating up from Mexico, where life is often even tougher. Over the last couple of weeks of harvest, I've gotten to know a few better; we exchange greetings, handshakes, and nods of the head. I like to think they appreciate my efforts to make their work tolerable. My fields have no toxic pesticide residues, and the green growth moderates temperatures. In addition to owning and managing these eighty acres of grapes and peaches, I make time to work with them in the fields, picking and stacking boxes, moving ladders and tractors. A Hispanic professor, a friend, once told me a story about his youth as a worker growing up in the fields of Salinas. He and others respected the Japanese farmers more than any one else because they always worked in the fields, side by side with the other workers. He felt that made a huge difference in how the workers were treated. I too spend time in the fields with my crew.

Do any of the men feel ashamed? I've seen photos of Issei (first-generation Japanese in America) dressed in their "Sunday best" for the photographer. They too were farm workers, poor and struggling. In their photographs, most wanted to project images of wealth and success. Or perhaps the photos in fact captured who they were on the inside: people of dignity. I can't believe any Issei actually worked in those black jackets with white shirts and dark ties or in the long skirts with hair tightly pulled back and held in place with some sort of jewelry pin. Their suits and dresses next to plow horses and barren fields seem only to accent the contrasts and challenges in their lives.

Or am I too quick to make judgments? Most of my workers support families in Mexico. During winter trips back home they are celebrated for their contributions to a family and the building of a good house, albeit in stages that take years to complete – a foundation with a good harvest one year, a roof with the next, and walls gradually rise as work in *el norte* is found. Many of the workers I hire do plan on returning home, to Mexico; for now, they can help their families only by sending money earned in these fields far away.

Photos of work. Are these men embarrassed by the label "menial labor"? Or is that a value others impose? I wonder if that is why my grandparents don't have photos of their farm work. I picture the faces that photographer Dorothea Lange captured so clearly – the hardened look of a Depression-era migrant worker, the lifeless eyes, the deeply cut wrinkles, the weathered face and matted hair. Maybe these images were the reason cameras had been kept away from the realities of the field.

One of the workers pauses and nods to me. He holds a huge peach in his dark hand, fingers curled around the luscious fruit while it still hangs on a branch. He lifts it toward the camera and smiles as I take a picture. The blush red of the fruit, the rich greens of the leaves, a smile in the dark face. He wants me to feel proud – a rich image of the fruits of my labor and his role in its harvest.

My camera rapidly clicks and adjusts for the lack of light with an automatic flash, illuminating the dark shadows of tree interiors as the men reach to retrieve ripe peaches. I imagine the photographers of my grandparents' era, grappling with cameras that needed to be mounted, single flash units prepped. The process was time-consuming and expensive back then, and few had the disposable income for photographs of ordinary work. Yet for these workers today, photographs are still an unnecessary expense, requiring a camera, film, and developing. I save money by sending my film out for processing by mail – it's cheap and simple, but requires a permanent address some of these workers don't have.

One of the workers says something to me about *familia y fotografia*. I don't completely understand, but it triggers an idea. I hop into my truck, run inside our farmhouse, and return with my Polaroid camera. As the crew finish picking and load ladders onto the trailer, I ask if anyone wants to have his picture taken. At first they don't understand – I've already taken many photographs of them. Then I hold up the different camera, and they begin to comprehend my question.

I sense they understand that a Polaroid is simple; individual pictures can be taken and given out immediately. I've seen street vendors in the larger cities of Mexico and South America standing in front of a picturesque market square or fountain. Couples (not tourists) stop and have their photo taken, pay money, and have a memory of their Sunday afternoon together. The owner only needs a camera and he or she makes money from the pictures.

I snap a photo and hand it to Merced. He's taken aback, not sure what to do. He thinks I'm just having him hold the photo. Then I explain in my limited Spanish vocabulary, "*Para tu, para tu familia.*" A look of surprise, then a smile breaks out on his face. This is for him. He thanks me and begins fanning the picture, speeding the developing process.

Immediately a line forms and all the workers want their picture taken. They stand with a peach tree behind them; some strike a pose as if they're picking a peach, others stand rigid but beaming. One worker shakes his finger in front of his face, motioning me to stop, then he points to the side and asks me to go with him. He stands in front of my truck and wants his photo taken there. I laugh, and the other workers start teasing him. They're giving him a bad time about the pose, and he just smiles. I grin and think of stories I've heard of Issei men taking a photograph of themselves in front of a large farmhouse and insinuating to prospective brides in Japan that this is their farm. I can imagine a carefully worded letter back to Japan with the picture — "I am farming the land you see behind me," or "The fields here in America are rich beyond belief." After I take the picture, the worker blurts, *"Para novia"* — For my sweetheart — and the crew roars with understanding laughter.

The next man in line stops and asks if he can take a picture with more than one.

"Simon!" I blurt, becoming braver with my limited Spanish.

He calls three others and four men stand shoulder to shoulder.

"Hermanos?" — Brothers? I ask, and they all nod in unison. I take two photos and make a request: *"Para familia en México, okay?"*

They nod, and one says, "We all brothers and want to send this home to *madre.*"

I think of my photo arriving in a village in rural Mexico. A mother has not seen her four sons for months, perhaps years. She places the image on the altar with the Virgin Mary, lights four candles, and prays for the safety of her sons.

My imagination may be running wild with the exchange, but I sense a sort of gratitude. Perhaps because my harvest is over and I won't see this crew until next year. For a few months this summer they have worked hard and helped me. I want to give something back to them beyond wages. I believe we all share something in common —

an understanding of working-class families. Perhaps my pictures help bridge two languages, two countries, two cultures, and two classes. Family pictures of a different sort, photographs for generations to come. Like a family story passed down, kept for years — pictures about work in America on a Japanese-American peach farm, images that return the soul to these laborers. For a brief moment on a summer morning, these men are visible.

I don't have a copy of any of these photographs. I have no negative for reprints. The snapshot is in my hands for only a moment. But the moment does not belong to me.

One younger man loiters. He's quiet and speaks with a heavy accent, a pidgin Spanish mixed with some sort of dialect. He's darker-skinned than the others. He asks if I can take another photograph. I can't understand his explanation, but it doesn't matter. This time he picks a big, red, juicy peach and cradles it in his hands, holding it chest-high. He grins. I take the picture, and he thanks me two or three times. "*De nada*," I answer. He runs to a waiting car, where a door is swung open, and he hops in. I can see him holding the Polaroid outside the window, flapping it in the breeze.

To See
the Future

I IMAGINE SOMETIME during the early 1900s, a farm family packed up their old car and left behind old farm tools and worn shovels. Behind them the farmhouse door slammed for a final time, and ahead lay the city and their future. They never looked back. From that point on, the majority of Americans were no longer rural. Since then, they've been running from the country into the city in droves.

My parents grew up in the following era, watching their friends and peers leave the farm. The Great Depression accelerated the pace, followed by wartime economies that fed jobs to the urban centers. The migration turned into a stampede, the exit into an exodus. As I grew up in the 1950s and 1960s, farming became efficient, demanding larger economies of scale and consolidation of farm land. Profit margins tightened, surpluses increased, and I remember hearing the phrase "There are just too many farmers." In 1900, one out of ten Americans operated a farm; by 1997, it was less than one in a hundred. That's why I was shocked to discover people wanted to hear stories from a farmer.

Hauling a shovel into a room full of teachers when you're not a teacher can be intimidating. Farm tools are not part of most curricula in California. Add "nonunion" to my identity and I might as well have come from Hayseed, Mars. Meanwhile the teachers' conversations bounced from SAT-9 to CAT tests, reading recovery to cuts in bilingual/multicultural programs to something called "lit-con." Ultimately sides were taken, some enthusiastic for new pedagogical programs while others launched vicious attacks against administrators. Afraid to take a side, I felt like an alien in a foreign land.

At first, I interpreted some of their comments as whining, but the chatter sounded an awful lot like the times farmers gather to complain about the weather or prices, kick at dirt clods over the topic of government regulations, and shrug about the future. All of my prior experience with teachers was from the viewpoint of the end user. Most teachers were dedicated to the old-fashioned and romantic notion of feeding the mind of a child, making a difference. They complained in order to vent, and I kept reminding myself: how would I feel spending five days a week with twenty or thirty kids? Or how would I sound after just one week of talking for hours with nine-year-olds? (My own speech deteriorates quickly after a single afternoon and one-sided debates with thirty-year-old peach trees.)

I've delivered keynote addresses to groups of foreign language teachers, high school English instructors, and educators involved with the Agriculture in the Classroom program. Most were women, and they reminded me of farm wives: the fiber that holds an operation together. While single-minded husbands tackled the simple-minded jobs like driving a tractor straight down a row and into poverty, the farm wives kept track of expenses and income, labor and taxes.

Add the jobs of raising children, volunteering in school, heading church and community groups, all in a rural setting with limited resources. Teachers did the grunt work of education and were on the front lines, *mano a mano* with our children and their families.

I started my talk by taking the teachers away from their classrooms and into my world on a farm. I described my wonderful peaches that had nearly become homeless because they didn't suit the demands of fast farming, fast turnover, fast profits and results. Heads nodded in agreement.

"Farmers and teachers are not factory workers. Good peaches and good students must be grown slow. Trees and young minds are certainly not machines." I sounded like a radical at a rally at Berkeley. I loved the feeling.

"Just as there's an art to farming, there's an art to teaching. Our work is not about speed and productivity. It includes a necessary slowness." I spoke deliberately. Of course, this was not new to the teachers, but I sensed they were surprised by my reminders. The concept of slow in education had been equated to dumbing down. With the recent emphasis on test scores and state and federal standards, a school year had begun to resemble a race. Standardized tests don't measure slow, contemplative thinking.

"Fast-food education doesn't work. Learning is not an exact science with a single method appropriate for everyone. Education must be slow because it's a human act – filled with wonderful individual subjectivity and expression. There are no master training manuals for your classes, and teachers are not management trainees. You are planted in a room, must grow on the job, and quickly become the expert – a process repeated anew each year. Young teachers and farmers rely on a youthful energy; older ones draw on a reservoir of experience.

"Farmers and teachers work with similar timelines. Our profession embodies this perspective: we see the past in the present in order to shape the future. Each peach tree or grapevine or student carries baggage from the past. Our challenge is to help shape it for the future."

I picked up my shovel and boldly held it up. I pointed to the tip, worn by use, and shared my story about generations working with this shovel – slicing weeds, turning the earth, digging ditches. I moved my finger along an imaginary line where the metal once was and explained, "For my grandparents and parents, it took years to hone this metal."

I continued, "Shovels should be the visual timeline that each student carries from grade to grade. Each year, a teacher helps sharpen the blade. It requires years to wear a half inch off the metal. Decades later, a fine, honed tool may result. Learning is the art of gradual accrual."

Before, I had pictured education as being like a trip to a grocery store. You load goods from the different aisles into your shopping cart, picking through the fresh new math methods, stocking up on the three R's, splurging on alternative summer sessions, and treating yourself to desserts of creative elective courses. A good consumer selects the right mix, beginning with parents' choosing the proper elementary and high schools and interviewing teachers for the job of educating their children. College may be a mutual decision of family and student, but individual majors and classes are dictated by young adults, pushing the cart by themselves for the first time.

Yet from a teacher's viewpoint, perhaps education is like metal sculpting: each instructor filing away, cutting and welding, adding experiences not products, expanding skills instead of inputting data.

"Learning is the sharpening of who and what we are. Education blends the senses – to see the future with the touch of our teachers. Just as the blade of my old worn shovel reflects the work of those before me and a family farm bears the marks of generations on the land, each of us carries within us the art of our teachers."

Walking Without Thoreau

I AM NO THOREAU when it comes to walking. He recommended spending four hours a day "sauntering" through the woods, never in a hurry, "ruminating" while strolling. For me, initially it's uncomfortable. Walking means to quit work, to force myself not to run, to go slow. Who has the time? My first steps prompt questions. Where am I going to? What am I looking for? What's the purpose?

Walking, I can't cover a lot of ground. There's not much work I can accomplish. I'm often lost without the proper tool if I see something that needs to be done. I can't establish a good pace; my legs are poor equipment in my farm's race to harvest. I fear I'll lose track of time, falling farther behind in my chores.

But I force myself to walk, rationalizing that it helps me monitor my fields. Still, I find myself walking with a fixation on a destination (imagining one even when none exists). I uncover more jobs to do, jot notes to myself, making endless lists of chores I'll never completely finish. In a compromise to modern farming, my excursions have evolved into short, slow walks, though when accompanied by Thoreau, I also carry guilt for hurrying this time meant for rumination.

In order to slow down, I succumb to an impulse to dig weeds. Like an anxious smoker who needs something to steady his nervous hands. Instead of a cigarette, I carry a shovel. The tool helps me engage, stretch my muscles, study the fields as I work, appeasing the farmer part of me while the artist still scoffs, "Bring nothing." I wonder what naturalist and author Annie Dillard took on her walks at Tinker Creek.

Old farmers are better at walking. I don't think they have agendas. They say, "Nothing else to do," as they set out in the morning to lose themselves in the fields. I watch them trudge down a row, then slip into the greenery, enveloped by their vineyards or orchards. Hours later they return, empty-handed, with dust on their eyelashes and fresh grass stains on their shovel blades. I long for their rhythm.

The group of farmers and students huddles on the edge of the field in a slight chill from early morning, spring dew fairly light and quickly drying. They tap their feet, shove their hands in their pants pockets, cross and uncross their arms and look around. They want to get into the fields and be told what to look for by the entomologist who's supposed to lead the farm walk. It's a brilliant tactic to make us wait. Impatience brings out the creative anticipation in some; others fill the void with dull chatter as if we were in a shopping mall.

"What do you see?" the organizer suddenly asks.

I see the wide avenue between fields, twenty feet, perhaps more. No, this is a dividing line between farms. I notice that each farmer disks differently. One side drove a tandem disk with what appears to be four gangs of blades slicing and turning weeds and earth and leaving a seam cut into the dirt; the other side, probably a drag disk or spring tooth that levels the ground smooth. I notice the tracks from the tractors, tires biting into the soft dirt, slipping sideways as they whipped around and headed back into their respective fields. I think of the rice fields in Japan and the maze of narrow earthen trails atop the low dikes, not quite roads, although vehicles and carts have left ruts behind. The paths wind their way through the landscape, respecting the lay of the land, adjusting to the contours of the gentle slopes, and the family histories of ownership that must accompany each curve and twist.

Once, my neighbor Jim and I stared down our shared avenue, unsure where the line separates our properties. Jim used a shovel as a temporary marker defining what may have been our boundary line. The stake looked silly, standing naked, alone, as if it could mark some invisible line between neighbors. I shrugged. He shrugged and tossed the shovel into the back of his pickup. Later he yanked out the end vines from each of his rows to make a wider border. Jim had purchased a big diesel tractor and a wide hydraulic drag disk; they required a broad open area to turn around and made close encounters with my vines. Jim respected our shared avenue and sacrificed his vines.

A friend who worked in town as an arts administrator for a non-profit organization owned a small parcel of land and told me of an encounter with his rural neighbor. They had met for the first time over a barbed-wire fence that separated the ranchette from the rancher and his cattle. They exchanged greetings, a little history, then parted after the rancher said, "Tell you what. You take care of your side of the fence and I'll take care of mine. We'll get along just fine that way." My friend thought for days about the power of that metaphor – taking care of your side of the fence. A perfect image of tolerance.

A month later he spied his neighbor's truck by the fence and stopped. "Don't know what broke the fence here. I found it first, so I'll fix it." Old hands searched through a toolbox for the right crimping tool as two severed strands of wire were reattached, pulled taut with a grunt. "You find any breaks on your side lately?"

My friend rethought his metaphor and replaced it with a more literal meaning. "Taking care of your side of the fence" meant, well, taking care of your side of the fence. Discovering a break and fixing it. Mending the boundary made a simple formula for getting along.

The group grows anxious and wants to surge into the orchard. I delay as long as I can, trying to picture this farm from overhead, seeing it in context of neighbors. The farm tour leader steps forward into the field; most of the farmers and students bolt and follow his lead. The entomologist tells us to take our time. "Observe and watch. Not looking for anything specific. Do this without expectations." A few farmers slow and take a deep breath; others seem to sigh, fighting impatience.

I pace along the avenue. A farm walk doesn't need to be linear, and I'm not sure a circle works either. Perhaps going in a spiral – but how do I manage three dimensions and depth? That's not possible unless I bob and weave, raising and lowering my head to observe multiple planes. I crouch down low. A grid pattern of brown tree trunks dominates. Twenty feet by twenty feet, an old spacing design where trees spread their limbs far and wide. Newer orchards are planted denser, more trees per acre and higher productivity. Eventually, though, the branches will grow outward and shade the same ground.

"An acre of sunshine is an acre of sunshine no matter how close trees are planted," an old-timer once told me. But closer plantings do achieve quicker results – by the fourth and fifth years, more trees produce more fruit per acre. Only when full-grown, by their eighth, ninth, and tenth years when branches reach out wider, can old-fashioned plantings catch up. By then, some newer variety will have replaced the old and forced the farmer into a replanting decision. But if you keep orchards decades, it really doesn't matter much. I squat and try to see the timelines used on this land – this farmer appears to have patience.

A visiting artist once told me she felt uncomfortable in my fields. "The grid pattern of the trees, the rigid order of rows and lines, don't feel natural at all," she explained. "The orchard is full of prisoners."

She was correct. My fields are not natural; they were planted by my family in straight rows and precise patterns. But no tree grew the same, no two seasons were ever alike. I was bothered and hurt and took her comments personally.

Most of the group has slipped deep into the orchard as I loiter behind. I crouch on all fours, then put my cheek down to the ground and look sideways at the coarse brown trunks. The subtle twists and turns each tell a story. I pan up several trees to where their canopies grow into a chaotic mass of green. Creative tension – an uneven balance of nature and human hands, wildness and precision.

The first steps are painfully slow, explains the entomologist. He barks out some instructions: "Study the density of green, watch the orchard floor canopy, feel the habitat for life." Some have gathered around him; he looks both annoyed and flattered. I take a step or two in, eyes adjusting to the shadows, then step back into the sunlight. Is this how insects see, shades and tones of light? Panning the field, I can identify the familiar: trees, limbs, leaves, grasses, weeds, bare soil, wildflowers. I know I'm biased and at first see what I expect. I quickly want to assign names, limiting what I see to what I can identify.

I gingerly step forward. Sunlight glares through some limbs, while others are dense and dark. Do some trees have fewer leaves because they're weak and sick, or am I looking at a young immature branch? A decade ago a fertilizer salesman showed me a weak section of my orchard and persuaded me to purchase tons and tons of a proper fertilizer mix, NPK, for the entire farm. Convinced by my eyes, I believed him, but the trees were weak only in that corner. Another time, I had a leaf analysis conducted and was told my fields were in the middle of the chart, my trees were low in certain elements and could use nutrient inputs. They advised me what to add, and I added accordingly. Later I wondered – was each tree lacking or were half the trees missing something while the other half of the field was just fine? Savvy salespeople show us how to see what they want us to — nurturing our perceptions so that we happily follow their line of sight.

Sunlight and the seasons, the green of spring and winter? I think of my whimsical "Masumoto Theory" of being able to see the sky in order to judge good pruning. I'm not interested in the branches but rather the gaps between the branches — the negative space. Only later did I realize I was probably trying to see, in the dead of winter and leafless trees, how sunlight needed to blanket all the trees as much as possible, and seeing the sky helps measure the openness of a tree. The negative spaces become the best indicator of how much light can penetrate and induce the magic of photosynthesis. I can see the summer sunlight in the gray, overcast fog of our winters.

We walk with our backs to the morning sun shining over shoulders. The small peaches glow. From a distance they look larger than they are, capturing the early angle of light. Closer, they aren't so big. Later in the morning, they shrink even more, washed out by the intense overhead sun. In the evening, especially along the western edge of the field where the late-afternoon sun strokes the golf-ball-sized fruit, they will have grown again in my mind's eye. I'm trained to see objects, not light — the sun plays with me, enlarging and diminishing what I expect to see.

I spend the entire day relearning how to walk, breaking habits, going much slower than feels comfortable. Some of the group stay into the afternoon, and a few of us return in the evening and walk into the sunset, no longer worried about searching for insect life.

In companies with "walking-around management," supervisors and bosses make a practice of strolling through their departments in order to maintain a hands-on, motivating contact with employees and their work. When the Gallo brothers were young, they'd come and walk the fields before buying grapes from a farmer. Instead of simply checking a few vines near the roadside, both brothers would walk, almost run, up and down rows, surveying the entire vineyard, assessing the quality before making an offer. They shrewdly saw what they were buying.

When visitors drop by and we talk in the fields, I can't help myself – as we chat, I'll start to poke at something green with my boots, scrapping and kicking until I finally reach down and rip out a weed. When the family takes a farm walk to get the mail or visit a neighbor, a ten-minute stroll turns into thirty as I pull one weed and then another. Marcy says it's a farmer's disease – you can't keep still, you're obsessed with work. I am compulsive about my weeds. I weed by hand.

Every year, a different weed seems to emerge and dominate my fields. Early in spring I need to determine if I can live with this new threat. Someone called this "just in time" management. I think of it as letting go and guessing. Once I ignored a tiny little plant called mare's tail. For the first week, the small clusters of leaves looked like marigolds, low to the ground, miniature green umbrellas gently pushing upward. Then they kept expanding their shade to about four or five inches wide and up, up, and up. By midsummer they were pushing through the vineyard leaves five feet in the air. They naturally reach for sunlight and didn't stop until my fields were dotted with green poles sticking through the tops of canopies, the main stalk hardening into a woody fiber, until finally, at seven feet tall, they pushed out flower and seed pods, preparing for next year's winter rains. It was too late to contain them, though I spent hours hand-pulling as many of them as I could (I had grown thousands), jerking the towers down through the canopy, occasionally plucking off grapes along with the thick stalks, shredding juicy bunches in the path of my angry yanks.

But other weeds I can live with. The dark green wild Italian ryegrass shimmers and sways in the spring breeze. These clumps spread quickly and sparkle in the sunlight. One city uncle claims my fields look rough, like the golf courses he frequents. I take comfort in the lushness, finding a home in the natural ground cover. I tell my uncle, "They're not weeds. I call them 'indigenous grasses.'" He shakes his head, and I know he's thinking about fairways and his battles with "indigenous grasses."

Chickweed spreads in early spring. Fed by late-winter showers, it circles the trunks like delicate ivy, hugging the bark low to the ground. With the last fogs of the season, tiny white flowers appear, the first blooms of the year. I can't slice and toss clumps to the side; the tangled mass frustrates shovels, wrapping around blades, handles, and trunks, refusing to let go. After we began farming organically, I often fell behind in work, and partly because it didn't look vicious, chickweed dropped low on my chore list. (Low-growing weeds often hide from a farmer's wary eye.) During farm walks I was still bothered to see it spread, gnawing at my vision of what a clean farm was supposed to look like. Then, with the first heat wave in May, it began to wither. At first I assumed it was the residual traces of herbicide still in the ground, yet each spring the grass turned yellow as if given a shot of weed killer. Gradually, I took credit for nature's handiwork, and now I leave the plants alone.

If my farm walks provide me with a diagnostic survey of the farm, what about those things I don't want to see? I can ignore chickweed, but what about brown rot in peaches or Pierce's disease in grapes? I don't see these until fruits begin to decay or vines decline and die.

In June and throughout the summer, I'll start lifting vine canes in order to examine grape bunches, searching for signs of a white, powdery coating of mildew on the berries and stems. Mildew and other diseases such as bunch rot and botrytis stealthily invade a vineyard, anchor themselves, and grow, infecting tissue and scarring the surfaces, stunting berries and cracking the skins. (With certain grapes at certain times of the year, among vintners botrytis is welcomed and even nurtured. I've had some wonderful wine from botrytis grapes — but not from my grapes, which are meant for raisins, not wine.)

To control the mildew, I dust a light coat of sulfur, and in ideal spring weather (between seventy and ninety degrees) the fumes of the vaporizing sulfur become toxic to fungus. But other diseases are much harder to control using organic farming methods. We don't have an arsenal of fungicides at our disposal. So throughout the late spring and summer, I'll walk and check vine after vine, hoping I will not discover an outbreak. By midsummer the vine canopies will have grown into a heavy curtain. I have to reposition canes and toss them over the vine, like lifting a veil, just to inspect the hanging bunches.

After weeks of monitoring, I announce to myself that the bunches look clean. Satisfied, I watch leaves gently sway under the dense layers of growth. That's when I recognize a possible reason my vineyards remain fairly disease-free: by opening the canopies and exposing the bunches to air, I can prevent the humid, dark microclimate conducive to diseases. A single leaf dancing with a subtle breeze helps make the air currents visible.

Over the course of years, I have re-trellised my fields so the vine canes are raised and lifted up, away from the hanging bunches, allowing air to circulate. Now, many farmers have gone one step further: they pull off leaves surrounding each bunch, so that grapes are exposed. I can't observe diseases when the first spores parachute into my fields. I can't watch them infect tissue or colonize blossoms. But I can pause and notice a leaf stirring, see the wind and trust it.

What else can't I see as I walk? I wonder about the amount of organic matter in my soils or the density of habitat for biological control. Is my soil creating ground vegetative cover and natural homes for good bugs? I have my own Masumoto measurement scale: healthy soils are not always seen but can be smelled — a rich, earthy aroma with a fragrance I want to breathe in deeply. Good earth can also be heard in the evening when critters come out and begin their nightly song. In the end I abandon trying to measure what I see, strip away my thoughts, and take a few playful strides.

How do you see things when you don't know what you're look-ing for? I've concluded that for pests farmers can observe — big weeds and fat worms — we've devised elaborate and very visible means to control them. Big tractors and multigang sets of disk blades leave a battered and torn trail behind; weed stalks and stems are not only sliced and ripped from the earth but turned over and buried in shal-low graves. Sprays kill instantly, within minutes. Years ago, when we used these toxins, sometimes I'd get off the tractor to check the sprayer and evaluate coverage. Before the mist had dried, dead bodies were dropping off leaves. Even deep inside the leaf canopies where I was sure the liquid had not penetrated, the fumes were effective and potent; instantaneous results were a measurement of success.

But many farmers blindly shoot chemicals at those pests not eas-ily seen — plant diseases and viruses, and even low-growing weeds or tiny larvae. Control is reduced to preemptive strikes early in the sea-son. If you see shot hole, brown rot, mildew, or phomopsis it's already too late, the damage has been done. Little research has been con-ducted to identify the invisible indicators and symptoms at their dif-ferent stages of infection and monitor their life cycles for possible intervention. Absence implies success.

As I walk, I often wish I carried a hand lens, to see things close up and in detail. (I tried keeping one in my shirt pocket but kept los-ing it when leaning over to pull a weed; I tried tucking one into my pants pocket and broke several when in lifting something heavy I'd rest the weight against my hips and thighs and hear the crack of glass or plastic.) Instead, I have learned a trick — I can sometimes see things by standing still. I rely on my peripheral vision, staring at a small branch or vine cane, not looking for anything in particular. That's when I see movement.

A healthy population of California gray ants reside in my fields. These are relatively large creatures, very timid, and they don't bite. I see them crawling in the peach branches and even the tops of my trees which stand over twelve feet high or scurrying over the twisted black trunks of vines. Their colonies are deep in the root zones of older trees and grapes. Tapping on the trunk with a shovel sends them into a frenzy; the ground comes alive with motion.

I teamed with a group of University of California scientists who documented that these ants were not simply strolling in my orchards, but foraging. Foraging? As in hunting? They sought food, especially a delicacy — the soft and fat larvae of a worm called the peach twig borer.

In an amazing sequence, the scientists showed me how ants handled a large worm. When one or two of the workers couldn't dislodge their prey, the ants called for reinforcements and a team returned to yank and pull until something gave way. The band of hunters then proudly marched home, carrying various pieces of their catch back to the colony, a fine dinner for their queen and her royal court.

The research project documented that without highly toxic pesticides and herbicides, gray ant populations surged and remained high throughout the growing season. Furthermore, in my old trees with aging roots and dead sections of woody tissue, we found large, healthy colonies. Old-growth trunks seemed to provide an excellent habitat for the ants.

I now walk slowly, hoping to discover my allies busy at work, helping to protect my crop of peaches. I have to train my eyes; it's hard to not seek an object and look for nothing. When my fourteen-year-old daughter, Nikiko, and I walk, I try to teach her this. Her quickness sometimes dooms her. Once I asked, "Do you see the ants?" and I watched her eyes dart from branch to branch, racing to isolate the subject using her speed and youth.

I explained the goal was to see things "nearby," what I call "next-to searching." I then described a wonderful afternoon I once spent at the university library in Berkeley, roaming the quiet floors, amazed by the vast knowledge all around me. The scent of old books was enchanting

— a mix of leather, canvas binding, and old stories, like a visit into an old farmhouse attic and steamer trunk. I walked the canyons of stacks, a timeline of learning through the ages. I wandered and got lost, turned in a circle and stalked, escaping up a flight of stairs only to discover another floor and then another. Occasionally I sighted someone, standing with book in hand or seated in a dark wooden chair. Such people looked up with pale face and blank expression, lost in their work, in the company of the thousands of voices clamoring to be heard as a book was opened with the soft crack of a spine. Or were they ghosts?

Arriving in the section for rural sociology, I ran my fingers along the labels, looking for a match to the call letters and numbers jotted on a note card. Close to my destination, I crooked my neck and looked sideways, reading the titles, pausing to pull a few books. My thoughts were tugged in new directions. The publication I sought was missing, but never mind, those works "next to" and "nearby" had already affected my thinking. Ideas filled me. I was still in a discovery stage of research, expanding and stretching my thinking when appropriate and necessary.

Nikiko listened and shrugged. She had very, very little experience walking a library. "That's not how we research on the Net," she calmly informed me.

Walking slow feels like going in circles. I stumble from tree to tree, sure I visited this one before, but I know better — the bend in the limb can't be duplicated, the pattern of leaves is unique, the cluster of tiny developing peaches is never the same. I easily see the ants on every tree now and begin to notice the hues of green — fresh pale leaves, older maturer darker leaves, young branches with vigor, older limbs with age. I see a *Sunday in the Park* with Georges Seurat; tiny individual dots blur together, the greens with the brown limbs and gray ants, and create something anew.

Farming must be circular, in contrast to the straight lines of business. If so, then family farming adds extra layers of concentric circles. I gradually make my rounds. See more, feel more, hear more. Imagination runs wild. I walk with voices, a morning passage from Thoreau, a midday stroll with Annie Dillard, the evenings with family and neighbors. I imagine my grandfather's quiet tone, a Japanese accent, his few words penetrating the accompanying pauses and silence between thoughts.

Haiku poetry must come from walking – the sound of a frog jumping into an old pond – a farmer walking slow, the taste of old peaches. I can see the twisted trunks and in a few months limbs heavy with fruit hang low to the ground. A flicker makes its home in an old tree trunk. I first spy a small pile of fresh red wood chips beneath an opening, marking the entryway of a new resident in my orchard. My noisy approach has scared the creature long before my arrival; I see it only through what was left behind.

I tell visitors how I regenerate these thirty-year-old peach trees. As an old limb dies, I guide a fresh shoot into its place. I run my fingers along the trunk, searching for seams where the old and new wood join, and I invite my guest to do the same. I can feel generations in a single tree.

I invite a small band of sixth-graders to the farm to share with them the stories and hope they learn through their senses. I point out the tiny, overlooked flowers of shepherd's purse, one of the first to bloom in my spring. One bright girl picks a handful, tickles her nose with the delicate blooms, and makes a bouquet to take home. She explains, "These are all about love, right? Look at the tiny heart-shaped leaves!" A detail I hadn't noticed before.

I add all these stories to my walk because that is how I remember. I learn by tales passed down, experienced myself, and told and retold. Stories call attention to what I see; stories create memory.

I step back out of the orchard into the open avenue. The bright sunlight glares, and I squeeze my eyes open and shut, hoping to adjust.

I reach for a shovel, throw it over my shoulders, and head home. Walking will continue another day.

Marcy says I look like Dad when I walk the fields. We both wear JCPenney blue work shirts, faded and patched jeans, dusty boots. "You two walk the same," she explains, and demonstrates – a slight slouch, shoulders round, leaning forward, more of a trudge than a stride. Perhaps I am now becoming a slow farmer.

The fields surround me, I see with my senses, aware. Is this what a monk feels while walking? Or the wise doctor, looking for symptoms but quieting the judging mind and instead open to signals nature is sending. Or the veteran athlete – feeling the field, reading a ball, hearing the sound of a swing, reaching deep within for performance.

I can imagine juicy and fat peaches, the red blush glowing in the late-afternoon sun, a tree with peach lights in it, a siren of harvest time. I'll be mesmerized, alive with flavor. My mouth salivates at the thought, I can almost breathe in the aroma.

PART TWO

*The Art
of Listening*

Things Silent

Dawn. Foggy mornings. Blistering midday sun. Day's end. Nights thinking about work. First blooms. Swelling buds. Ripening peaches. Grapes hanging fat. Golden fall leaves. Old farmers working slow. Between strides as you walk. When you stop breathing after a long deep sigh. Holding your breath while a dust devil swirls around you. As a child, end of day, after a hot bath in the *ofuro*. Standing outside in the cool evening air. Gazing up at the stars.

A fallen vine. Broken tree limb. Dust settling. Pathogens – molds, fungi, virus. Bulldozed field after its work is done. A gap I've only heard in the countryside, the pause high above in between a jet engine's noise as if they're shifting gears. The moment between lightning and thunder. An Arctic cold front with deadly frost. Summer heat radiating up from the earth. Crickets when they rest a measure. Wind without leaves. Ripples in a still pond spreading in concentric circles after you toss a rock.

Coming
Home

Fᴿᴏᴍ ᴏᴜʀ ᴏʀᴄʜᴀʀᴅs and vineyards, my parents saved enough to send three children to college. As profits shrank, the family goal to provide a better life for us kids meant leaving the farm. We didn't trust a future of working the land. Our story didn't seem to be worth much. Dad suggested, "Maybe you're better off if you don't farm," and his voice trailed off in defeat.

Why have I returned? At college in Berkeley, I was lost in the speed of the city and the crowds, traffic and pace, though at times the energy was invigorating, a welcome change from life in the Central Valley. The emotional baggage I carried from growing up on a farm seemed out of place along the teeming streets. I longed for silence, and I continued to rise early each day — comforted by the stillness of dawn, but even as I walked at sunrise along Telegraph Avenue in Berkeley, the stapled political posters flapping in the slightest breeze shouted out loud. I stepped over and on discarded leaflets that still echoed with sharp voices.

Then during my two years in Japan as an exchange student, I came to terms with a sense of obligation to my parents. I wanted to come home and take care of my folks while they were still healthy. I had questions to ask them about working the land before we began using chemicals and pesticides. (Our family farmed organically not by choice but because we were poor and our farm was small.) Natural farming requires a lengthy learning curve, built on knowledge native to a place. I wanted to "talk story" like the Hawaiians during balmy evenings or their relaxed and lazy afternoons when they exchanged story after story, singing and dancing with their words and images, pidgin mixed with dramatic gestures and intonation. I realized that my family answers would need to come forth over weeks, months, even years. It required that I make myself available by settling into a daily routine of talking "slow story."

Stories taught me how to farm. I sat in on conversations between my father and my uncles. They told of droughts, rain on raisins, heat waves in spring, trading tales of disasters and their feeble attempts to save a crop. I took notes.

Mom would stir her memory — a description of "lost birthdays" because she was born during the raisin harvest or a recollection of working silently next to a mother-in-law or lifting hundred-pound raisin sweat boxes with a partner husband. The words invited questions that gave license to reveal and laugh. Through oral histories I pieced together a mental map of the family farm landscape.

Neighbors dropped by to talk of farm disks and sprayers, of irrigation schedules and working sandy ground, of harvest expenses and leaving fruit hanging on trees with bad prices. "Break winter's crust in early spring." "You need a good head of water to reach the ends of quarter-mile rows." "Cut your losses and don't look down at the over-ripe fruit dropping from trees." I gleaned lessons from many lifetimes of experiences. I learned to trust these words.

I documented times and places and descriptions of field practices. I read scientific reports and studied charts, then conversed with researchers and absorbed their expertise. Farming seemed to have

more questions than clear, definitive answers. I enjoyed the ponder-
ing much more than problem-solving. Reflection felt natural, replica-
tion proved stubborn.

I began to write in journals and gradually collect stories. Vines
and trees dangled the lure of an official "off" season, and during their
dormancy I planned to create manuscripts. Initially, I envisioned I
could earn enough income from farming to support the hungry life as
a writer. I was wrong. I managed to find solace in the compatibility of
farming and writing, though neither made money. I was free to take
my time with both.

I was sometimes asked to speak at a Nisei neighbor's funeral. The best
eulogies were simply stories — notes of family histories, tales of
dreams and hope, a sharing of values and spirit. In order to write
these, I interviewed surviving spouses, and our conversations resulted
in wonderful oral histories. Once, I stopped a widow after hearing a
deeply moving, personal tale of their family farm and the late farmer's
ties to the land — emotions rarely expressed nor seen by the family.
After a pause, she whispered, "This place . . . it was his baby."

I stopped and said, "I know your children, they've left the farm.
But you shouldn't be telling me these stories. You should tell your
children. They need to know."

Her eyes were glassy; she rubbed together her rough, dry hands,
which had been quietly folded in her lap. Then she quickly answered,
"Aren't you trying to be a writer?" Her hands stopped. "You tell them
for me." Then she gently smiled; I swear it looked more like a grin.

I returned to the farm to hear stories. The widow's words became
inspiration and simultaneously a "burden of tradition" one generation
passes on to another. Here on the farm, voices from the past live in
the present.

On the farm, I have discovered a sense of place, somewhere I could start a family. Perhaps I had misinterpreted the pain of my father's work and did not understand his silence. After each winter, he still longed to return to his fields. He could lose himself in the tedious work and at the end of a day sense progress. For hours he'd labor in solitude, enveloped by the stillness of the farm. This was home. Now, with luck, good prices, and the help of old orchards and vineyards, I can work alone. Content with the quiet. I too may send my children to college – although I recently gave my fifteen-year-old daughter a quick lecture about the economics of farming; our "old peach scholarship fund" will probably suffice only for public, not private college.

Our old trees and vines may be badly bent, yet they keep producing abundantly like a stubborn old farmer who keeps working. Every harvest we're treated to luscious flavors, a taste that perseveres. My job is not to build and expand but sustain; to find a balance between the art of working the land and the economics of farming. Just as a story varies slightly with each telling, every year's weather leaves an imprint in my peaches and grapes – not all fruits need to taste exactly the same. I like to think mine have a balanced character that's alive.

Sound
of Work

Eﾠ ᴀʀʟʏ ᴀᴜᴛᴜᴍɴ morning, the farmer heard strange sounds. Muffled voices travel only short distances through the dense orange groves, so Jake Itogawa, a family friend and farmer, trudged toward the avenue dividing his land and his neighbor's. The hard red clay of the Sierra foothills differs greatly from the valley floor's sandy loam; citrus thrives along these rolling hills, picturesque but a nightmare to farm. With a heavy rain, the land wants to return to the pre-farmer era and run down natural gullies and miniature gorges. Orange trees planted in a manicured grid pattern stand in the way.

The voices mixed with a clanging of metal, followed by laughter and more chatter. Jake felt something was different about their tone, lighter and almost playful. If they were kids, the high-pitched chatter would soon be mixed with some yelps and hollering; every kid succumbs to the temptation to try and yell across the valley, somehow believing the Coast Ranges eighty miles away will echo back. But these were the deep voices of adults.

Stepping out of the trees, he found his neighbor's sons, all three of them, whom Jake thought of like nephews. He had watched them grow up in the groves, their father dragging them out into the fields to work. He remembers overhearing the boys' loud arguments with their strict father, a man "from the old country" as Jake called him, not much different than his own Issei father. The boys didn't want to work, but their father demanded they "pull their weight" around the ranch. During these verbal battles, Jake, working along the shared avenue, could hear the voices grow in intensity with curses and quick exchanges followed by prolonged silence, or else the crash of tools and equipment being tossed into the pickup with thuds and clangs, then a slamming of a door and an engine roaring, tires spinning and screeching. Jake imagined the father standing in the yard, dirt and dust kicked up in his face. Later the boys would return to another shouting match with their father.

As Jake approached, he saw the sons were clearing out a trench along the lane, a precaution on this day after Thanksgiving before the heavy winter storms marched in. The shallow ditch they were making was designed to catch water and gently guide it along the road, steering it into the wider, open flatlands below without taking out a hillside of oranges or washing out an avenue. The sons looked up and smiled. They stopped and shook with hearty handshakes, the moment no different from hundreds of other times when the boys had looked for an excuse to stop work and take a break, pausing with a neighbor. "I offered them the use of my tractor," Jake would later explain. "Could clear out that channel in just a few minutes, but they insisted on doing it by hand with their shovels." Jake shook his head. "Hardheaded, just like their dad."

The sons continued for hours, accompanied with farm talk that ebbed and flowed. Conversations rose above the thrusts of a shovel, increased in intensity, then slowed into a syncopated rhythm, the metal blades clanging together when the brothers worked the same area. The sons were working the place where the lane curved and a natural slope of the land normally catapulted water over the edge. The noise quieted for an hour, then quickly increased in tempo and volume in anticipation of quitting time.

They tossed the shovels into the truck bed, the handles crashing against the metal, the steel bouncing against walls and toolboxes. They opened and closed their hands, sore from the hard work. Blisters were swelling, some popped and shredded. Three grown sons come home to do some chores; two were doctors and the other a successful businessman. Their ailing father stayed at home but could hear them if the wind blew in the right direction.

One son explained to Jake: "Thanks for your offer to help. But we need to do this work." With a grin he adds, "Dad says it keeps us grounded."

Another son says, "Yeah, and we all know about being grounded."

The three brothers squeezed into an old Chevy truck. The engine roared and tires spun, then grabbed hold of the clay avenue with a squeal and a cab full of laughter.

Farmers spend hours and hours by themselves. We often work alone. We still talk a lot, mainly to ourselves – muttering at a tractor or disk; conversing with a vine we're pruning or a tree we're sawing; talking through problems and solutions with a farm dog (I don't think farm cats make particularly good listeners – they wander away right in the middle of my best soliloquies). We complement our solitude with our own sounds.

When we do meet each other, our voices are used to talking across a field, a type of yelling but not in anger, more like a greeting. Kato-san, a hardworking farmer with a lively character, immigrated from Japan as part of a farmer exchange/training program by the Japanese government to help cope with the depressed post–World War II rural agrarian sector. He stayed in our community and brought with him a love of sake, Japanese whiskey. His passion for this beverage seemed to accompany his verbal expressions, even in the middle of a plum orchard. A number of times I'd stop by the house and find his wife inside or working out in the back. She'd point to his general location and then from across the way, sometimes as much as a quarter mile, I'd hear a sharp, piercing voice.

"Ahhhh, Masumoto-san," he'd yell, "Ahhh, baaaad year. Noooo good. Noooo good." He'd smack his lips together with a sucking sound as if he'd just had a drink of his homeland. All that was missing was a deep, refreshing sigh.

We always began our conversations by exchanging self-deprecating remarks about how bad farming was and how stupid we were to still be farming. Even close up and face to face, Kato-san's voice remained the same, loud, biting, and punctuated with occasional smacking. If the topic evolved into something more positive, like talk about family and our children, he'd add a few drawn-out moans with a grin. With serious discussion, he'd purse his lips, and I knew he would love to "talk about it" over a drink. But *sake* at ten in the morning was too early for me.

Most farm sounds normally do not involve conversation. Nature is noisy enough. I can watch a summer sunset and listen to it echo over the countryside. A metal roof heated by the day's sun contracts and pops as the temperature cools. Trees seem to sigh and the vines take a deep breath with the breeze that originates from a drop in temperature. "Sundowner" winds, created by the rapid change in temperature when the sun sets and cooler air rolls off the nearby foothills, stir the farm into a frenzy, with leaves slapping and whipping about as the day prepares for the evening. Birds take evening flights, flapping overhead, in search of food and water. Other critters stir, like rabbits and insects, with the blast of direct heat dissipating.

Farm sounds increase with familiarity. Visitors often have an unsettled feeling about dawn, before the sunlight exposes the outline of the day, or dusk, when the remnants of light can illuminate. They don't trust their listening skills, even when I suggest they see by hearing. Darkness is foreign, and they have the urge to fill the visual void with chatter, uncomfortable with a moment of silence as if it implies failure.

At certain times of the year, I can estimate the time of day by sounds. Time not measured by clocks, but by a farmer's work pace. On a cold winter morning, I want to delay going out until it warms a little – I often wait until I can hear the light dusting of frost melting and the trees dripping before I head out. My timepiece – the sounds of nature – informs me that the biting cold is now gone.

I combine an art of listening with my daily chores, trying to sort out a rhythm. Specific jobs have a specific cadence if I pay attention to their sounds and my movements. Shoveling weeds has its own beat according to soil types and weeds. In hard ground, a shovel blade "shu-shu-shu's" along, scraping the surface in search of shallow roots. Deeper-rooted weeds need a slower "choo-choo-choo" work song as the blade penetrates and severs. Established weeds require a deep cut; I listen to my boot striking the top of the blade with a thump and forcing the steel cutting edge into the earth, a single smooth "slide and slice" downward, then a quick jerk and lift, flinging evil roots out into the open.

My performance art of weeding includes a complex series of movements and motions in conjunction with my hearing. I labor according to the acoustics of the shovel, changing grips and angles as a bed of rooted weeds announces itself with a thud that sends a vibration through the wooden handle or the penetration of heavier soil that slows the tempo and stroke diminuendo. I make hundreds of adjustments, most very subtle and without a lot of thought. I spy the shallow-rooted chickweed and drop the handle low, flattening my stroke as the blade swims just below the surface. Sometimes I begin to daydream, yet the shovel continues to twist and whirl, and as I enter a pocket of moist ground, the coarse "shuu-ing" gives way to a softer "choo-ing" and I shorten my stroke, adjusting to the change. I don't miss a beat; my work rhythm flows and it all sounds good to me.

Listening isn't limited to hand labor. When I disk one vineyard, a strip of hardpan sits just below the surface along an eastern edge. A passing disk will change key; the smooth slicing of dirt gives way to a moment of grinding steel against rock. I cringe; a shiver crawls up my back as if fingernails were scratching a chalkboard. Over the years I've mentally flagged this spot, anticipating the rock iceberg. But even though the vines are laid out in a grid pattern with twelve-foot rows and each vine seven feet apart, the hardpan strip doesn't run at a right angle to the vines. It sits diagonally, a natural angle. My hearing wants it placed in a pattern I can recognize so that next time I won't be surprised. I grow uncomfortable not knowing the exact location of the sound.

Farm
Animals

HAVING GROWN UP on a goat dairy and having raised the occasional steer and sheep, Marcy shares stories about the sounds of her family farm – especially the milking hours listening to the radio for *Polka Party with Dick Sinclair* harmonized with the noise of farm animals – the cry of hunger or the bawl of wanting to be milked; the screams of fear and the howl of sexual impulses. Marcy claims many creatures are quieter when happy. Then the only sounds are the soft chewing of cud like the purring of a cat.

When visiting my in-laws, I remember the wail of a goat at night when an infant kid was separated from its mother. I swore it was a human baby crying. Piercing, it woke me. I looked out the window, fearing what I might see – an abandoned babe or a haunting ghost? A voice from *The Silence of the Lambs*. Outside the bedroom window stood a small herd of their goats – someone must have left the pen gate open.

On our farm we have a few chickens, and their clucking doesn't seem to fit any pattern. I claim they're not that smart, and Nikiko and Korio don't like my rude generalization. The creatures scream once in a while when laying an egg. Even though the shell's surface is smooth and seems the proper shape to slip out from a hen, I can hear a cackle, cluk, cluk, than a louder cackle, cluk, cluk, and finally a scream with a cry of desperation, "push, push, push." *Voilà* – a pretty big egg from our rather small Rhode Island Reds and the Araucana.

Occasionally a bossy hen chases the others, ending in a flutter of wings and a chatter of complaints. Then the flock returns to clucking, a guttural sound not even close to singing. We also had a rooster who, for years, always crowed at sunrise. His behavior wasn't quite that consistent though – remember, he was still part of the chicken family – because after crowing with the first peak of sun over the Sierras, he would also occasionally crow throughout the day and even at night for no good reason other than to perplex us. (I eventually stopped trying to find a pattern. Who was more confused, him or me?)

Sounds at night can baffle. Once while working late in the barn, I heard scurrying behind one of the few walls that had Sheetrock (which I learned was left over from a farmhouse remodeling project, because no farmer would plan on Sheetrocking an old barn). As I tightened some plow bolts, I could hear a gnawing sound from behind the wall. I tried to isolate the location, knocking with my wrench at the spot, sending tiny feet dashing in different directions. A few minutes later, the munching resumed.

I knew food was being dragged to the safety of a mouse dining room, tucked behind a wall, secure from predators. I've often heard the sound of walnuts rolled to a new picnic location when the old one was discovered. But this was different: the creature kept returning to the same spot no matter how much I knocked, tapped, invaded its space.

I marked the spot with chalk, and a few weeks later, in the cold of winter, I heard it again. The same grating and scraping sound. I grew more than curious, alarmed this time, because hungry mice will gnaw at electrical wires. They have eaten the bushing off my irrigation pump, which I discovered when I needed water during a spring drought. They've even gotten under my car hood at night, attracted first to the warmth of a cooling engine, to feast on colorful plastic cables. I've heard them scrambling there in early morning, a soft thumping, then a fleshy tearing sound from the engine area; I opted not to inspect the damage until later.

Fearful of my barn's fragile electrical system, I tore open the Sheetrock to discover the source of the mice's winter feast. I found the shriveled, leathery body of an ancient cat who must have crawled into the wall some time ago and died. Over the years the skin hardened with the dry heat of our summers, and the lack of rain and humidity had cured it. In desperate times, the mice munched on this cat in a twisted ballad of role reversal.

Without animal sounds, a farm can become eerily quiet. I never realized how much I missed the noise from critters until our old dog Jake died. He had given up barking in his old age. Poor eyesight and a loss of hearing didn't help his ability to detect strangers, and lulled with long hours of deep sleep, he seemed quiet and content. When he passed away, though, I missed our morning routine. When I'd step out for work, Jake would slowly rise after a momentary debate if I was worth the effort. Then he'd stretch with a slow-motion tai-chi move and finally shake his head. His ears flapped as he shook his body awake. The flopping sound became the perfect way to start the day, a calm yet quick tempo, a gentle and persistent beat, a natural sound I wish alarm clocks had. As I've grown older, I wish I could shake out the stiffness of my body like Jake and flop my ears to waken my senses to a new day.

In the distance I can hear the pounding of metal, a hammer forcefully descending and crashing into something solid. In a different time, it might have been the clang of a blacksmith and his anvil, but I cannot say for sure, since I have never heard a blacksmith at work. I do remember as a child driving by an old, old wooden building in Del Rey when the sign read "Blacksmith," yet I cannot recall a memory of the sound.

In our old barn I study the horse stalls and shutters that opened to the outside. The wood is abraded along the bottom wall of the windows, gnawed by animals and worn by outstretched necks and a daily rubbing against the edge. Along one wall hang assorted tools supported by rusting nails. I try to imagine their sounds when utilized – the "shhh-ick, shhh-ick" of sheep shears, the shrill grind of a sharpening-stone wheel, the "twang" of a huge spring that must have supported some piece of harvesting equipment.

I'm most intrigued with the weathered harnesses and the series of "T's," wooden bars with steel rings attached, used when hitching a team of work animals to something that needs to be pulled. I can imagine the snapping of leather and the wood creaking under the weight as a wagon or plow is yanked along. The harnesses rub against flesh, the metal rings jingle as the team picks up speed. In my imagination, eventually a human voice cries out. Movielike slang echo. "Giddy-up." "Yeee-ha." And a clicking noise from the corner of a mouth with the crack of the reins.

Reality snaps into my daydream. When he was a younger, Dad farmed with animals and may have used these tools. Our family was poor and tractors were for the rich. We had mules named Jackie and Molly, and they needed verbal prompting from the farmer. I have never thought of my quiet and reserved father cajoling a team of work animals. I smile when I think of him shouting "Yeee-ha," then realize he probably used a more Japanese-sounding cry. "Hiii-ya?" Next a chant, a howl and chatter, a bark and call – "Com'n, com'n. Haiii! Haiii! Haiii! C'mon, Jackie 'n' Molly. C'mon, old Jackie 'n' Molly." Dad's work song to encourage his animals and himself. A quiet chant, the song of plowing.

Working with my seventy-nine-year-old father, I can hear his joints crack, his breathing labored at times, his sighs of exertion. I can't hear his internal workings, though, a racing heart, aching muscles, blood surging through veins, a deadly blood clot dislodged and winding through his system. When he wants to do some work, I now try to find a lighter job without hard, physical strain. I'm concerned he'll labor too hard and my assigned tasks will kill him. Yet were he to die, he'd want to go while out in the fields.

Periodically, I try to check on him in an unobtrusive manner. I know he'd quickly detect if I were observing him and resent my distrust. So I listen for him instead. Sometimes I can hear his pruning shears snipping at vine canes or the sound of a shovel scraping against a concrete irrigation value. Other times the roar of a tractor tells me where he's working; as the tempo drops I know he's at the end of a row, and when the engine accelerates, I know he's completed the turn safely and has started a new one.

I enjoy finding him the most by listening for his pocket transistor radio. When he's tuned to a baseball game, I stop and pause, straining my senses, trying to pick up the sound of the announcer, the buzz of the crowd, the crack of the bat. I turn, trying to find the best reception, just like twisting the small AM radio at odd angles to locate a stronger signal. Dad has taken our family to only a few baseball parks in our life, but we were at thousands of games while working together and tuning in the Giants or Dodgers and later the A's and Angels. A farmer's day at the park. Beginning with the optimism of spring training and opening day, through the long, dog days of summer and night games (we were often still out in the fields for the first few innings), and finally with the change of season accompanying the fall classic, announcers and players kept us company. We almost always paused with a crucial play, listening to hear if a ball sailed over a fence – "Bye, bye, baby!" or "And you can tell it goodbye!" We'd tighten our work hands into fists with a close game or clench our teeth with a pitcher-

and-batter duel; a threat of a no-hitter made our knuckles white, and we'd end the day exhausted.

Today, I don't believe Dad follows his teams as closely with the ballooning of salaries and the business of baseball dominating the game. Yet I can't help a childlike grin when I can still locate him in the fields, guided by the voice of Vin Scully or Jon Miller as he helps us farmers see the players, smell the grass, feel the tension, and hear the sounds of the game.

Night Work

I KNOW I WILL NOT finish in time. Light fading as I head west, I watch the setting sun cast a lavender glow with oranges and yellows gathering above the Coast Ranges. The sun seems to hover for a bit just above the low mountain range, then once an edge dips below the horizon, it drops quickly, sinking like a deflating balloon. I'm sure it's an optical illusion, but it also signals the start of darkening light. I know that after that moment I can no longer use my vision and start relying on my hearing.

Shadows darken my path. As I drive down the center of the vine rows, the vine trellises on each side are my guides. I'm unconcerned as I lose sight of the disk blades behind me. From the very beginning of the day I lost them in the churning cloud of dust and instead trusted the feel of the tractor, sensing a drag under the strain if something lodged in the blades that prevented them from rotating. Gradually the darkness robs me of sight. I'll try switching on the tractor lights, but they rarely work. Bouncing down mile after mile of fields, wires quickly wiggle loose. I couldn't remember the last time I bothered to check if they were functioning.

The phrase "darkness falling" is incorrect. The term implies something sudden, as if one moment you can see, then it "falls," and the next moment you're blind. Instead, darkness gradually settles over a landscape. I'm enveloped, a cold hug of darkness. Sunlight fades, vision drains. We're strangers to darkness. My straining to see further blinds me; as when driving in fog, staring longer and harder at a spot does not help. I am searching for familiar objects, seeing by identifying things relative to the direction I want to travel. I look for shadows. Initially the vines three feet on each side of the tractor provide guideposts. Slowly they melt into the dark and I realize I have to pay attention to something else.

Echolocation – the guidance system of bats, modified for old tractors with broken headlights. As I churn down a row, sightless, my steering falters and I start sailing off line, sliding to one side or the other. Earlier in the evening, I could still see the outline of vines swinging closer and closer to me. But now a canopy of leaves creates a wall, the engine sounds bounce back, and as I drift to one side, the volume increases and crescendos into a roar, warning me to steer clear.

The limits of echolocation have to do with speed. I can't drive fast, because correcting a course requires a gradual redirection in order to steer my ship back on line. A sudden roar and panic of discovering a vine inches away and the tractor yanks in the opposite direction, an overcompensation that generates an erratic swaying and weaving. I begin to jerk the vehicle into a fishtailing swing, rocking left to right, fixed only by slowing down. Sudden stopping at the wrong time will also send the heavy gang of disks staggering wide left or right into a vine trunk; coupled with forward momentum, the metal blade will slice the wood, gouge the bark, and rip off a hunk, a tearing sound that will echo in my mind for hours.

So at one point in the darkness I slow, listening to the engine churning at a seemingly safe distance from the vines. I can detect another sound, a scraping of vine leaves against the tractor body. Vigorous vines will extend canes across a row, and tendrils interlock like two lovers reaching over a forbidden territory. I don't pay attention to them until suddenly one cane grabs the tractor throttle on the dashboard, jerking it down, revving the engine and sending us flying forward. Adrenaline shoots into my system. I lunge to throttle down, but the tendril hooks my wrist, a cold touch against my skin sending chills down my back and across my chest. I feel as if evil hands are yanking me off the tractor. I try to dodge; they slap and grab. I crouch in my seat, hiding, hoping they miss. The black forest of night, the prince of darkness, wraps a cold finger around my flesh, wanting me. I whip around and scream in the night, "Haaa-ya!" while mimicking a weak karate chop, snapping the tendril in two, breaking the cold clutch. I can hear my heart racing, "tha-thump, tha-thump."

Another cane rubs the tractor and gently slides over my shoulder and arm, and I feel a bit foolish. I close my eyes, trying to refocus on the engine sounds, but a fat cane scratches the sheet metal with a shrill shriek. The next paws the metal body with leaves sliding along the frame. One thick cane gently rubs the surface, then bounces into the night air. I imagine it floating upward before dropping and stroking my cheek. I shiver with the companionship of massaging vines. In the moonlight, I steer clear of shadows and trust the sounds to guide me.

Farmers used to irrigate at night. A generation ago, most farms didn't have electric pumps and had to water during their turn with the runoff from the Sierra snow pack. A forty-acre farm might have only a few precious days and nights to cover the entire land. Miss your turn and the vines and trees would die. Farmers became night irrigators, guiding water down furrows, nudging it up to high ground, scouting for breaks in lines and flooded banks. Morning came with a true sense of accomplishment, a reward for an all-night vigil, and the vines sparkled in the early dew, satisfied with water.

I don't work with such pressure, but I still irrigate many nights, pleading for water to journey down a long quarter-mile row and reach the end by morning. All of my vines are connected to a cement piping system that runs along one end of each row. I can spread the water by opening and closing valves. The challenge is to allocate the individual trickle equally over a five-to-ten-acre block of vines, then move on to the next block. I have to check and recheck flows, fine-tuning the little gates on each valve, regulating flow so I can sleep for a few hours at night. By my final inspection before I retire, I've often lost all sunlight. A flashlight provides poor illumination; some irrigation valves are hidden by a canopy of leaves. Instead I rely on the sound of the water spilling out of the gates from a dozen valves, listening to the liquid tumbling out of a small two-inch hole in the side of a cement pipe, cascading down to the earth a few inches below. The sounds vary, and some are muffled by the collection of green grass and weeds growing at the base of each valve. Imagination befriends my efforts as I walk along the avenue. Initially I hear little, but as I listen for the silence, the trickle of water grows louder.

During the months of July and August, the Japanese-American community celebrates the Obon season, an annual festival honoring ancestors with lively street dancing and colorful dress. We believe that the spirits of our family come visit us each summer, initially attracted to the bright colors and upbeat folk songs. I like to think they come for the best peaches of the year, and as the last of my fruits are harvested in early August, the ancestors return to their heavenly resting places during another special ceremony called Toronagashi.

This event is held near water, a stream or slow river; in Fresno, with a lack of much flowing water, a small lake plays host to the event. We gather in the late evening to launch small boats, each with a lit candle and the name of a deceased family member written on a sheet of paper. One by one each craft is placed in the water and with a gentle nudge, an ancestor is directed back to his or her resting place. Even though the lake is still, usually a gentle breeze guides the fleet away from the shore and into the darkness of the lake. Only the candles reveal their direction.

While each family launches their ship, the name of the ancestor is read. Often a family recalls someone recently passed away as a voice echoes in the still evening air. I hear names of farmers now gone with no one to take over the family place. A name, a face, a memory. I'm reminded of the reading of names aloud on the Vietnam Memorial when it was dedicated. A starkly simple fact, the power of names being read.

Toronagashi helps close my fruit harvest. Each season a memorial to the past. I think of the names of the old fruit varieties we've grown but pulled out decades ago, varieties now grown by only very few farmers. Fay Elberta. J. H. Hale. Red Haven. Rio Oso Gem. Le Grand nectarines. Cleaning up after a long day of work, I once found in an old tin coffee can, tucked on a high shelf in the barn, the rubber stamps with these names, hand stamps we used to ink and press onto wooden boxes packed with these flavors. Names I read aloud as the evening gave way to darkness.

Songs of
a Shovel

I HAD NEVER WALKED in five-degree cold with a shovel in hand. The winter had swept in, a dry, Arctic mass that cracked fingertips, chafed cheeks, and chapped lips. I was unprepared. Our normal California winters may dip into the high twenties with lots of fog and heavy dews; the cold isn't biting, dissipating by midday to the thirties and forties, but the wet chill stays with you and your clothes.

But this freeze was compounded by insecurity as I walked the streets of Manhattan from a taxi caught in traffic, making my way through the everyday throng of commuters on sidewalks, weaving my way toward the Crowne Plaza on Broadway. I would soon speak at the annual conference sponsored by Chamber Music of America, the national umbrella organization of chamber musicians, presenters, and managers.

I was the farmer in the big city, feeling lost and insignificant, armed with trusty shovel and prepared for probing stares. But my farm tool garnered little attention, at least that I could notice. People don't make eye contact on the streets of New York; they walk with a hardened look of determination, a race to reach a destination, propelled faster by the intense cold. The only glances seemed to be from other outsiders, visitors like me who have never owned a dark overcoat, and our colorful light jackets looked out of place on the gray winter streets of New York. Black was still in.

Usually, farmers are not invited to New York to speak and chamber musicians do not gather to hear stories from farmers. But the conference planners took a chance on bringing in a fresh voice and I took a crash course on chamber music. The last time I played my French horn was in the eighth grade, but I gambled that those who created good music also enjoy good food. Perhaps a wonderful peach story could be as pleasant as a lively minuet or jaunty American fiddle music.

With shovel in hand, I began: "We share much in common." I had learned of problems chamber musicians know well in the transportation of their musical instruments. Violins and flutes travel easily, but cellos, basses, and large horns create tension-filled flights. "I too had great difficulty transporting the tools of my trade. Passing through check-in was unsettling. Metal detectors are not designed for farmers and musicians. And I don't even want to repeat my heated discussion about the size of my carry-on." With scattered chuckles and smiles I began to feel comfortable with this audience.

"We share an affinity in the practice of our work. In my fields, I rehearse for months with harvest as my ultimate performance. I work with a valued instrument – my shovel – not as a utilitarian tool but as a valued companion. I perform with an emotion of caring, but getting the notes right isn't enough. I struggle with the art of making art and the craft of making a living.

"We face a similar paradox because a chamber group, like a family farm, is not just a business. We're all emotionally involved, personally engaged and spiritually invested. We perform despite limited economic incentives because we offer gifts to the world.

"We survive as both art and business, and I draw the lines

between the two with our instruments. Shovels don't belong on factory farms, fine chamber music may never adequately compete with the mass entertainment market of sports, movies, or rock and roll. Instead, we're forced to live and work in a balancing act."

I was inspired by this collection of artists. They loved music and sought almost any means necessary to make it. A very few made their living solely from recordings and concert tours; most augmented their music earnings by teaching. All had loyal supporters — family and friends, endowed groups and their grants. Family farming too had adopted a similar survival strategy in relying on outside income sources. I had wondered if a natural affinity didn't link farming with art because we started each day with an addictive passion to create and craft.

"Complex simplicity. We struggle with an unpredictable nature, never quite sure what's going to work, creating products without knowing if there's a market. And now we face new technologies that displace farmers with machines and musicians with computers and synthesizers. Yet we accept the chaos in our fields because in something wild lives the necessary spirit of life — the precious element for our art. Call it creative pandemonium.

"In our work we accept both simple and complex answers. Chamber music is unconducted. It thrives as a collection of individuals learning to work as a group. There's no leader, yet clearly there's leadership. You learn to work with cues from those around you, subtle and pronounced, unspoken yet loud."

I thought of the flavor in a good peach — a delicate balance of sugar and acid embodied within the yellow flesh. Not saccharine peaches, nor overly tart, but a movement of flavors, wonderfully complex.

I held up my shovel, describing the variety of uses on the farm, the simplicity of the tool to create opportunity to explore. "With my shovel, I can read landscapes, feel the subtle nuances of soils, assess the moisture of a field. This tool can quickly dispose of a weed, yet with the intention of returning, for there will always be more weeds. Shovels force long-term thinking and imply that I will be around for a while.

"Any system that claims to farm and grow real peaches without a shovel is going to fail by our standards. My shovel empowers me to look for complex answers and work with an artistic vision. Shovels imply there's no single correct answer just as there's no right way to play a composition. It's the human factor that brands our product different, the power of live performance in our fields. Our art is not about cloning perfection but interpretation touched by a human spirit."

I spent three days in New York surrounded by musicians, music, and fine food and dreamed of quartets stopping over on the farm between concert tours of San Francisco and Los Angeles. I'd invite my neighbors, farmers and their families, to enjoy music under a harvest full moon, ninety-year-old vines twenty feet away providing the proper acoustics with peach trees gently swaying in a summer breeze. We'd treat the ensembles to fresh peaches, plums, and nectarines with flavors only farm families know and a meal with homegrown vegetables and family recipes. The evening would be filled with stories of simple folk with a love for music — many neighbors grew up in rural churches with choirs or small schools with struggling music programs. We farmers knew of rhythm — the cadence of a shovel swinging back and forth, scraping the hard earth of summer; the knocking of a tractor engine, gauging the health of equipment; or the resonance of fruit pinched off and tumbling on the steps of worker's ladders as we thin peaches in spring.

I left the conference with new sounds turning in my thoughts, enough to fill another season. "The world will always need farmers and musicians to feed our bodies and souls. The music of my peaches honors the flavor of our shared spirit, and a farmer's hope is what I and my shovel leave with you today."

Farm Music

A bird does not sing because it has the answer, it sings because it has a song.

——CHINESE PROVERB

I own a Buddhist tractor. When it's running well, I hear the "oooom-mmm" of a finely tuned engine and the potential for great work. I believe that in the past, most farmers connected spiritual beliefs to farm sounds. I imagine a Jewish peasant talking out loud with animals, bargaining to complete his work, "Oh dear God, why do you test me? Can't you help my horse this one time?" Or a Midwestern dairy farmer and the sound of boots running into the kitchen, the splash of holy water resting above the sink, then a dash back to the manure spreader while praying for the gear drive to work and for the chains and belts to hold together on a freezing morning. I can envision a Mexican family flicking a match with a hiss in order to light another candle before a shrine in the kitchen, the Virgin Mary accompanied by other artifacts – a cross sent from California and a Polaroid of a distant son working in the fields – and a prayer that sounds like a chant for good harvests.

But many of these spiritual farm sounds and songs are lost in modern farming. I have never heard a folk song about planting, harvest, or change of season in California. I have no chant to bless a new tractor or plow (although the Armenians in the Fresno area delightfully still have a blessing of the grapes — I cheat and go to their ceremony, hoping their good luck will rub off). Few farmers talk to their trees or vines anymore (at least they are animate objects), let alone their tools and equipment. Farming in California is just over one hundred years old; perhaps that's not enough time for folk cultures to adapt to the changes. Today we have little spiritual work, we mostly do business.

That's why I hope to hear a Buddhist chant from my tractor. I usually try to develop a schedule for the day, a sequence of steps. Disk an orchard with thick weeds, then furrow young trees that need water soon, and finally return to the slow task of French plowing of a vineyard. By the time I reach the tractor, I have a list in my pocket and am anxious to get started. Much of this agenda vanishes when I turn the key and the starter clicks instead of turns. The sound is sharp and piercing, a slight jolt to the senses as I recall the weak battery or the finicky solenoid. Diesel engines have a different character from their gas cousins; they're harder to start, but once they get going, they don't need a strong electrical system. So I plead for the starter to turn over, promising I'll take care of it next time. "Please, please, please." Begging in such situations keeps me humble; getting angry has never helped. I pause, take a deep breath, delaying as I calculate the loss of time with a dead battery, wondering if I should switch to another tractor with less horsepower. Then, with a tentative twist, I turn the key. "Ka-roooooommm." The sound is sweet. A reminder to myself — celebrate the little things.

Initially, the diesel engine knocks and pounds, silencing other noises. The power thunders raw and massive, and I suddenly feel vulnerable. I grip the steering wheel tightly, recalling the problems I've had with too much force, the tendency to run too fast and wild and hook vine stakes or slam into tree trunks, bouncing off of them as I forge through obstacles. Driving down a path to my fields, the tractor begins to hum in a constant refrain, but as I bounce over potholes in the avenue, it sounds as though the engine misses and runs erratically.

Once in the fields, I drop the blades of the disk and start in low gear. The tractor settles into a constant pull with a slight drag on the engine, the sound softened by the leaves, grasses, and damp earth. I stop fighting the movement and listen closely. With a patch of choking weeds, extra-hard ground, or a soggy, waterlogged spot, the engine loses RPMs and strains. I respond by changing gears or momentarily lifting the disk to clear the obstruction. Gradually I slip into a rhythm. I anticipate tough terrain and can shift without stopping, coasting with an engaged clutch, switching gears smoothly. Almost uncon- sciously I throttle down whenever the engine begins to labor, adjust- ing when it begins to race. The diesel motor, even with its naturally knocking sounds, churns constantly through the different loads and terrains. I relax, no longer needing to struggle. The sound of a trac- tor, music to my ears.

The long, tedious hours of driving up and down row after row blur together as I become lost in my thoughts. I sometimes do my best writing and thinking on a tractor. The rattling reduces to a ringing harmony, the clamor subsides into a constant rhythm, a beautiful cacophony that blocks out all distractions.

Sound of a
Ripe Peach

Does a peach blossom pop like a poppyseed pod? We call the stage just before they open "popcorn" bloom, no doubt based on our visual, not our auditory, sense. But I must admit, I haven't put my ear up to one, nor have I sat for hours to watch. I have, though, heard the sound of a ripe peach.

At a peach's perfection, the juices swell and the aroma enchants; the skin blushes red and fiery orange. What's missing is the sounds. (Diane Ackerman in her wonderful *A Natural History of the Senses* claims that's why we clink glasses, because hearing is the only sense missing from full enjoyment of a wine). Yet I know of two types of ripe peach sounds. One is the noise of eating a peach: the breaking of the skin as teeth sink into the flesh, the sucking of juices out of the meat, the first noisy chews with mouth open as the nectars wash our taste buds, and the smacking of lips and tongue.

Farmers know of another peach sound, one that occurs in the private moment when a ripe peach falls from a tree. A peach drops, fat and overripe, because it has been overlooked or worse. Too many plops against the earth is the painful sound of low prices. We leave fruit to fall "splat" on the ground when there is so little demand that the costs of harvest are greater than our return. A shameful waste, man-made pain for all those involved, followed by a malicious taunting of the senses as the crop rots in the fields.

I try to block out such noises and hear only the plop of a ripe peach as a single moment of fullness. A piece of fruit so heavy and plump it can't support itself by the stem, its meat rich and lusty, tumbling to the ground and bursting upon impact. Brilliant orange fibers splash across the dirt, dripping and oozing, staining the soil with a luxurious decadence. Gobs of jam.

I smile and laugh at myself, thinking of a Japanese-American peach farmer's *haiku*:

> *Old farm*
> *Ripe peach leaps*
> *Kur plop.*

Story Songs

I CAN'T SING. Nor do I play a musical instrument with much skill — I plunk at a piano and pick at a guitar to amuse myself, then quickly grow frustrated at a lack of talent. (Although once when Nikiko was in sixth grade, I was asked to pick up a French horn I played in elementary school. At a holiday program, I had great fun accompanying her fifteen-piece country school band. We sounded terrible and wonderfully authentic.) Every Wednesday evening, a neighbor and his wife, a hardworking farm couple who spend long hours in the fields, join others at church and sing. They soar with hymns and I hear another side of their work overlooked in the dust and worms and lousy prices we too often talk about.

But I can read my stories aloud with emotion and passion while blending a lightness and conversational character with my images and scenes. I like the performance of a public reading, the connection with audience, the chemistry of a live presentation. Adding a voice to my farm stories creates a new layer of meaning, texture, and depth. Some of my writing is meant to be read aloud. I hope it reaches the power of poetry — words flung into the air, flying, towering, floating, drifting, sailing with an audience breathing in deeply.

With a talented musician friend, Larry, I try to blend music and the sound of farming through "story songs," combining my spoken words with his jazz, folk, and blues. We've experimented with a Japanese folk song on his flute as I describe my grandmother's farm hands – rough and thick with callus yet generating a warmth of massage. I hear myself as part of a long tradition of farmers and their tales of hard physical work.

Larry and I explore meaning by combining images with sounds of the familiar. I tell of my mother's attempt to create the ideal Thanksgiving table – she was a poor cook and tried to learn from magazines. Larry plays a rendition from *Appalachian Spring* on his clarinet. Our feast came with a distinct family flavor: we never ate cranberries with turkey or applesauce with ham because we believed they were not condiments but desserts. And Dad still had to have white rice with his stuffing. The Shaker hymn "Simple Gifts," completes the scene, the cadence of an unfolding American tradition: a Japanese-American family knew what the holiday table was supposed to look like, but no one told us how to eat it.

Larry's saxophone chants as I read a story about the death of my grandfather. The sound of a *shakuhachi*, airy with a dreamlike tone, joins my picture of Jiichan/Grandpa, clipping vine rootings, preparing them for planting and the birth of a new vineyard. When Jiichan slumped over, probably suffering from a stroke, my then four-year-old brother cried for Baachan/Grandma and she bolted to get her sons, who were preparing the fields. My brother ran into the house to get his fallen grandfather the only thing he could offer. The grandson tried to give his dying grandfather a drink of water. If this works right, the wandering blues of a saxophone combine with the story and the audience listens with a tingle on the back of their necks, a flush of blood over their faces, and a shiver up their spines. The emotional message connects with music, not as ballad nor as background, but as a soaring of spirit.

Nikiko plays the Japanese *taiko* drum and we perform a duet, she

on the bigger *taiko* while I read and play a smaller, h'
shime-daiko. A song of a hailstorm, written with the sou..
lent weather front racing into our valley, winds whipping, rain hu..
horizontal and the ripping razor edges of hailstones smashing and slit-
ting peach flesh. Our drums pound a rhythm, thunder rolls, bodies
fling, and arms flail. The tearing of the hail, a piercing, jarring whip
and beat, sounds, words, and movement in one powerful moment.
Vulnerable. We lost an entire crop in ten minutes. The black clouds
marched to a neighbor's and we could hear the same song repeated
over and over at another farm. At the end I am weary, our song almost
too real as I revisit the storm with my daughter.

I can hear many sounds at the same time, unlike sight and colors – mix
red and black and get something else; blend too many colors and end
up with puke gray. But we can listen to multiple sounds and separate
them. I hear two dogs barking, the deeper sound of a neighbor's lab
and the higher-pitched sound of a mutt. Engines from different
sources – one must be a tractor with a broken muffler, loud and
obnoxious; another a truck, lumbering along the road with gears
shifting; finally the deep whine of a Caterpillar, another vineyard
removed or an orchard pushed over? Overhead birds, actually a flock
of individual birds flapping and swooping, then a quick turn and they
split in two, one veering north and the other downward, fluttering
wings shift and twist. The best speaker system can't pick up all the
motion and the three-dimensional movement. Quickly they all land
on a wire and the rush is over; then they sing.

Farm sounds blend and mix and I add yet another layer, the element of time. I hear our vine pruners clip and snip, then yank on the
canes to pull them from the trellis. The wood snaps and the wire
squeals, scraping against the metal staples anchoring them in redwood
stakes. My father and his father heard the same work, and I can hear

generations laboring, the sounds similar, a related resonance, a music shared.

Farm sounds can carry memories that connect me with the past through even the simplest of noises. I can hear wood being cut by hand and think of the uneven strokes as Dad sawed off a branch, awkwardly pushing the thick limb away from the blade, helping gravity pull it downward. A hacking, grating rhythm. I picture the time Dad climbs into the tree and perches himself by straddling two other branches, precariously balanced, trying to be careful as he saws and forces the branch to split, the wood cracking under his weight. His pace quickens, the final strands snap and pop, I can hear a heart racing as arms pump back and forth, pulling teeth through the fibers. Haaack. Whhoook. Haaack. Whhoook. Haaack. With a nimble hop to a safe branch, he lets the log fall with a thud. Quiet. I can hear Dad's heavy breathing.

The Art of
Grunting

A DOCUMENTARY ABOUT African-American railroad workers contained a segment about a crew of men setting rail. They held long metal shafts in their hands, tips anchored against the side of a steel rail, and with a coordinated grunt, they'd push it into place, accompanied by another team hammering a spike in place. The chant and hammering percussion allowed the men to work for hours in unison, keeping pace as a team.

Our family had no such work songs or chants. I theorize it's because of the distance between peach trees or grape vines. Peaches are usually planted twenty feet apart and vines rows are twelve feet wide (seven feet between the individual vines in a single row). That means if one of us got slightly ahead of the others, even a single tree stagger meant shouting forty feet to hold a conversation. A dense canopy of leaves makes it hard to hear or see each other. (Low-growing crops like strawberries may promote a chorus of conversations instead of solos.) Even when we worked in pairs, I can't think of an organized arrangement of rhythms or a choreographed dance to help pass the time or add to teamwork rhythms. Some of our workers, especially the

Mexicans, have their own yelps and barks that come at all sorts of times — a cry of a coyote or the screech of a bird, mixed with some rude comment about another workmate's sister or mother, instigating random responses, chatter and laughter.

But like all manual laborers, we have our own pattern of grunting. These remain individual acts, random moments of expression. Grunting can be done solo or in duets, occasionally in a trio or more, but that's rare and difficult to coordinate.

Japanese are great grunters, men more so than women, farmers some of the best. I remember Issei men with very expressive grunts. The simplest form "ne" could be stretched out, "nnnneeeeee," to mean either deep thinking or a sigh of relief, all depending on intonation and a nod of the head. A rise in expression with a slight nod meant approval; a drop in inflection and the subtle shake was negative. The more sparing the turn of the head back and forth often meant the worse the scenario. Japanese usually kept public disclosures of deep pain to a minimum. Also, a monotone "neeee" implied some sort of thought was required or the Issei had no answers and most likely no clue what to say next; the proper response was then to quickly drop the subject rather than cause embarrassment.

The grunt "oh" usually lacked physical accompaniment, and only the most adept listeners could interpret it clearly. "Oh!" in a short, sudden burst could mean surprise, disagreement, displeasure, or deep concern. "Ohhh" with a quick ascending tone could communicate, well, surprise, disagreement, displeasure, or deep concern, but definitely something different from the flat version of "ohhh." Some employed a slightly longer form — "oh-rooo" — by adding an extra sound, doubling the complexity of meaning. The scariest form was the "ohhh" that drifted into a low and airy silence, like the wandering sound of a *shakuhachi*, the voice of a ghost with a message of doom. I learned to watch if the eyebrows tightened with this version of "ohhh" — which implied something wrong, almost dreadful, or an attempt to disguise the speaker's confusion. Either way, explanations rarely followed.

"Yo-i-sho" became the word of choice for almost all Japanese, no matter which *ken*, or province, they had immigrated from. "Yoi-sho" cut across class lines and geographic differences; city and farm folks, rich and poor, men and women repeated this expressive grunt.

"Yoisho" has been perfected by centuries of Japanese working in the fields and by hand. It has a natural sound, a cadence that dances to the beat of the work to be done. Monotonous chores are accompanied by a stretched version suited to a slower pace — "yooo — iiii — shoooo" — which sounds like a marching chant by a drill sergeant. Need to speed up the frequency? Shorten to "yooiii-shooo." Workers by the hour usually don't like this faster pace — piecework rates, payment by quantity not time, prompt this livelier beat. In the fields, one hears the class struggle: boss's grunts of "yooii-shooo" contrasting workers' grunts of "yooo-iii-shooo."

An even faster pace requires you to drop and slur syllables, "yoo-shoo, yoo-sho." I can sense a race brewing with this frequency, a competition between workers, a sibling rivalry, a generational test and rite of passage. One side forging ahead with a slightly quicker, perhaps more youthful "yoo-shoo," but halfway down a row or stack or pile, the sage and consistent "yoo-ii-shoo" catches and surpasses immature vigor. Most jobs reflect the grunts of distance runners, not sprinters. A few times, though, a killer speed is called for, usually reserved to a single burst of energy. It comes with a clear, definitive "yoosh," a decisive grunt and the potential for a hernia.

Unstacking and stacking boxes, bags, buckets, pick boxes, sweats, pallets, and piles, all lend themselves to "yoisho" variations. The grunt accompanies other senses; a rhythmic hand clap or slap is added to the physical nature of a task. Grab, lift, "yoooo-iii." Stack, push, position, "shooo." Followed by a double pat of the hand against the side of the box as if to signal it's done. Then the dance repeats. Stack, stack,

stack, pat, pat. With the final box, an ending coda of clap, clap as the next pallet is positioned and the stacking starts anew.

These phrases are not played with a measured beat or score. The art of grunting is random repetition, a cushion to comfort us and help us to endure. After all, this is called "grunt work."

Tractors and farm equipment can grunt too. I hook a shredder, an implement with spinning flails that whack, chop, snap, and splinter the branches and twigs from winter pruning of vines and trees. A violent machine, it gobbles up the piled brush that sits between the rows, chews and spits out the fibers and broken pieces of wood, ready to be disked into the earth. The work feels fast, gnawing row after row, power-takeoff shaft whirling, gears grinding, the drum spinning and churning, But I'm actually traveling very, very slowly; it takes hours to complete a few acres.

I can monitor the vigor of a field by the grunts of my equipment. When the tractor and shredder hit a pocket of very dense growth with lots of pruned clippings, the added volume grinds the flails to a near-halt. They struggle with the increased work, belts start to spin and smoke, the engine groans. I can fix this by stopping the tractor, dropping to a lower gear, and noting the location. In the next row I look for repeat patterns in certain sections of a field, drawing a mental map of my farm, studying the healthy trees along one edge to understand why something is working right as opposed to why something is wrong.

A research team from the University of California at Davis once discussed a statistical method of monitoring large blocks of farmlands. They used the term "geostatistics" to describe how they made measurements along a transit in order to gain a glimpse of the whole. I liked their approach to capturing the variations within a field, something many research projects don't examine. When I described my "grunt theory" to Bill, the research director, it brought to mind memories of his farm family's cornfields in Wisconsin. As their old harvester traveled down a row, he could see and hear the amount of corn being picked and shot into the trailer. He had his own "geostatistic" measurement, which was good enough for their operation. I'm sure with a good crop of corn, Bill and his father ended the harvest with a fulfilling grunt.

I am convinced there are dialects of grunts, reflecting a mix of ethnicity, a dash of spiritual beliefs, and a pinch of climate, though I have conducted only limited research.

The Midwest does have a "you betcha" style of conversation and grunting which is light and chirpy with a slurring of some vowels in a guttural voice. When some Midwestern farmers talk, they save energy by limiting the motion of their lips and mouth, not wasting energy on clear enunciation and storing up the complex words. Most daily exchanges necessarily include a repetition of the obvious: "You got a point there." It may have to do with the deep-rooted spiritual beliefs of Catholics or Lutherans with a little liberal United Church of Christ mixed in. I don't think a lot of folks in the territory of Marcy's Midwestern relatives were comfortable with expressing themselves publicly, especially their emotions. So grunting offers a wonderful opportunity to communicate and keep word count to a thrifty minimum.

One of Marcy's cousins, Bubba (really, that is his nickname), has perfected this art by combining the word "wow" with various grunting styles and idioms. He has successfully mastered dozens of expressions with a single word skillfully impregnated with a multitude of throaty sounds and sighs. He can convey surprise with a short and even "wow." A troubled "wow" lasts longer and is airy. Depression (as in getting a terrible hand in cards or breaking a train in dominos) is stretched and fades. After a few old-fashioneds, he can't disguise his grunts and blurts "wow" with a youthful glee as all the other domino players quickly start to dump points.

When Bubba and I talk farming and trade disaster stories, he responds with a "wow" I'm very familiar with — the grunt of disbelief and pain. Good news is greeted with an upbeat "wow," and the rare success story has me using his rhythm: we both delight in a heartfelt "wow" of joy and gratitude. I still have much to learn from Bubba; there are "wows" I haven't heard and grunts he may not even know he's capable of.

Silence of
a Clap

A LEADER HAD DIED, a saint some called him, the leader of farm workers and the poor. Tens of thousands gathered to mourn under tents and the spring evening sky in a dusty field in Delano, 1993. Cesar Chavez, the head of the United Farm Workers, had passed away, and I sat at his vigil, nights before his funeral. Speakers were honoring the man, thousands came to offer tribute, row after row, mostly poor, mostly Hispanic, mostly quiet. A radical priest spoke to the crowds in Spanish with emotion and passion, comparing Chavez with Christ, as a radical and a leader of the poor.

Then he started a clap. A single clap and the crowd quickly followed, most slightly late and off the beat. A long pause and another single clap. People stopped talking; children sat still and were captivated. Another clap, more in unison. The priest held his hands together, before spreading them widely, directing, leading, guiding. A clap that began to sound as one. The pauses became distinct. A clap and another, with an electricity circulating in the audience.

A clap and I thought of the thousands of hands. A clap and I thought of the hope Chavez brought to these people. A clap and the power the group felt, the noise they were making, the penned-up energy of people wanting to express themselves publicly. A clap and the sound that was no longer quiet.

The claps increased in speed, the pauses shorter and shorter. Everyone was pounding their hands, the noise deafening. The cadence increased, faster and faster, the pauses barely heard yet present with their own clarity. The pace sped even quicker, the beat remained tight, the people as one. Thunder shook the evening, hearts were racing. Finally it crossed over into rapid applause, the clapping random, the intensity broken. People then whistled and cheered, UFW flags were waved, voices cried out and shouted. No one wanted to stop.

The clapping still echoes within me. Years later, it's part of the sound of my farm. I'm still moved by the moment, but not by the noise or volume. The silence between the claps pierces deeply; the silence empowers the thunder of the clapping.

Farmers hear silence more than anything else. A silence of pain and disaster, laboring in isolation with low income. Yet a work filled with a quiet tranquillity, peaceful. I listen to our family farm and have my doubts. The challenges remain great. I have few solutions and often answer with silence. And that's okay.

I prune in the winter fog, the mist acting as a filter, I hide in the quiet. I step out in the early morning when the sun is just asserting itself, a moment of transformation between night and day. I trudge home late, a walking meditation. At first I can hear dogs barking, a truck grinding in the distance. Then my own boots crush a rock, my pant legs rub together, my own footsteps softly brush the earth. Finally, I listen to my own breathing, and if I'm lucky, I can hear my own heartbeat, and then nothing.

PART THREE

The Art of Taste

Memories of Taste

～

Juice. Peach, trickling down cheeks. Nectarines, dripping on your T-shirt. Oranges, hand-picked and chilled by morning air. Muscats and a tangy smacking of the lips before you spit out the seed. A single grape held between teeth, bursting.

Sweet. Meat of raisins rolling across your tongue. Last grapes of the year so heavy with natural sugars they make you thirsty. Plums as you suck out the flesh. Stringy fibers of overripe peaches, dangling between your teeth.

Grandpa and Grandma's backyard fruit tree. Picked for breakfast, gorging until it hurts. Summer snack of fruit you picked yourself. Youthful decadence, one bite of a soft, overripe tip of a peach and toss the rest. Flavors your parents know but you don't, and may never. Tastes you've tried and hope your children learn before it's too late. Memories grow richer with time and more real with age.

Got *Umami*?

U*MAMI* — savory, possibly the fifth taste, after the four conventional tastes: sweet, sour, bitter, and salty.

I had first heard of it while trying to define the Japanese term *shibui* — astringent. *Shibui* was used to describe the flavor of persimmon but also something "quiet and tasteful" — even some Japanese pottery can be *shibui*. I was thinking my farm could be *shibui*, but other meanings of the word, "glum" and "sour," didn't go well with my peaches and raisins. But could my farm have *umami*?

Food experts scoffed at the concept. "Taste and flavor are two separate things," they explained. Flavor is for chefs, artists, and writers (and I'd add farmers). Taste is for the tongue and identifiable receptors found in the taste buds.

I accepted their research and stopped my quest, trying to describe my peaches as sweet, but not like candy. Someone asked me about white-fleshed peaches, which are low in acids. I tried to explain that I like the tanginess of my fruits because they balance the sugars with natural acids. "So your peaches are bitter?" "No, green peaches are bitter — and sour."

Then I described the depth of flavor in a good raisin: "It has a type of aftertaste that stays with you, if you pay attention." I thought of a pain specialist whom journalist Bill Moyers once interviewed. Their discussion focused on a program for chronic pain sufferers — focus and meditation — and they began training by slowly chewing and eating a single raisin. "I lick my lips when I eat a raisin," I proclaimed. A food editor once asked if the natural sweetness of raisins also had a salty taste — that's why the combination had a depth of flavor. "No, it was more like a good wine." That opened the door for an outburst of wine talk, and my limited knowledge left me in the fields with my vines; I lacked the necessary vocabulary.

Some believed *umami* to be part of an absolute natural moment of perfection. The pristine strawberry lightly coated with spring morning dew, sparkling in the first morning light. I fondly dreamed of such a taste and searched each year for that single peach of perfection. I thought I found one and ate it, much to my delight, and then became overwhelmed with concern — "now what?" A muse then told me that I must not have truly found it.

I decided to return to the source and traveled to Japan to discover the original meaning of *umami*. In the early 1900s, Kikunae Ikeda of the Tokyo Imperial University was credited by some for the first discovery of a fifth taste. He was intrigued by the distinctive taste of seaweed and claimed to isolate at least one element of *umami* — the amino acid glutamate. This sounded like MSG, monosodium glutamate, a controversial flavor enhancer that sometimes gave me a rush, an artificial flavor high. Peaches with MSG? Not the organic roots I was seeking,

I met with a kindly university professor who specialized in food and *umami*. We met in the lobby of a large hotel where international business people and guests mingled. Over a ten-dollar slice of a melon (which was fantastic), we talked about food. I asked if peaches, specifically my organic peaches with their delayed tree-ripened harvest, had *umami*.

The professor smiled, shook his head no. I was crushed.

He explained that his interpretation of *umami* involved a natural process which altered the taste. His examples included fine cheeses and hams. "These foods undergo a change," he stated. "The process brings out the *umami*." He spoke of things fermented, like *miso* and wines. Then he added things dried. "We don't eat seaweed directly from the ocean. It has to dry, to cure."

"Like raisins?" I joked. He smiled. He nodded. I was redeemed.

I told him of my father's gourmet raisins. He waits until the workers have gone through the fields for harvest; afterward he walks up and down the rows, searching for the bunches they missed, a late gleaning weeks after the crop is normally in. He dries these in his yard, watching them daily, covering them at night to keep moisture away (so much for morning dew), turning each bunch so they dry evenly, then rolling and boxing each tray.

The professor grinned and said, "Humph. . . . *Hai* . . . yes, *umami*. . . ."

I arrived home content that at least my raisins had it. I realized the flavor of Dad's raisins were also affected by his presence – they were made sweeter because they were from his hands, the story part of the taste. So I add a human element to my definition of *umami*.

A year later, at a wine-tasting workshop, a maverick Master of Wine expert upset some in the audience. "We need to stop educating people about wine," he claimed, "and instead we need to listen to what they like." He blasted the traditional notions of matching wines with foods and advocated a novel approach, exploring what some people simply "like," then trying to determine why with a new classifying system. One element of this new coding method may be *umami*.

During a question-and-answer period, I asked if peaches have it. He said yes. I smiled. He explained that some foods were naturally high in *umami*, whereas others must undergo a change during their growing, ripening, or processing to enhance this taste, similar to the variability in aging, curing, or fermentation. We talked of ripeness and typical weather patterns, soil health and microclimates that may affect a peach. I certainly knew peaches, like grapes, in some years were outstanding, in others not as good.

Transformation. That's what makes a peach great and raisins wonderful. I can't identify all the elements and measure what it takes. I can't model a prediction as proof. I'm not completely sure about umami but like to think of it as a mystery. And the solution lies in the journey. I don't know if I've "got umami" but the fun becomes trying to find out. Perhaps that's why I keep farming – my search for a life I can savor.

Inspection

I LEARN FROM painful stories. In the mid-eighties, tree fruit prices tumbled, and no matter how hard I worked, peaches did not sell well. Even the prices of raisins tumbled because of an oversupply and declining sales. To augment our income, I took a part-time job with the USDA, inspecting raisins delivered by other farmers to a processing plant where the crop was to be washed and packed into boxes for shipping and sales. We had had a cool, wet spring and mild summer. The grapes never matured well, and the quality was terrible, with low sugars and berries that were more skin than juice. The individual raisins lacked plumpness and looked anemic and anorexic. Some farmers suffered more than others, depending on their soil type and farming skills.

In late October, near the end of the season, a final load was delivered. My job was to collect a representative sample from the truckload of forty-eight bins (each bin approximately a thousand pounds). My test bag would be taken into an on-site lab where machines measured the weight and quality of the raisins. While waiting for the truck to be unloaded, a fellow inspector and I wandered over to the first bin. We scanned the harvest, ran our fingers over the dried berries, unconsciously grabbing a few, snapping off stems and popping them into our mouths. We unofficially sampled a load, snacking and chewing the plump morsels, tasting the sweet meat – a treat of another harvest.

But this lot was terrible, the quality lousy. The shriveled skins were hard, with no meat inside. Good-quality raisins have fine wrinkles; these had a few deep gouges with particles of sand trapped in them. My teeth crunched on the grit, and the grinding vibration sent a shudder through me. I wanted to spit out this crop; it had the flavor of failure.

In order to fill my collection bag, I had to stick my hand deep into each raisin bin. The individual berries were so hard they jabbed my fingertips and tore at my cuticles. The stems were coarse and scratching; one hooked my knuckle and ripped the skin back. I thought of the pathetic vineyard these must have come from, weak vines and an inexperienced farmer. I contended that the entire load would never pass.

I hadn't noticed before, but off to the side stood a small white-haired man. He held his arms close to his body and slouched, as if huddled in a corner. He kept blinking, and his face was tight. He appeared to be biting his lower lip, sucking on it, and occasionally made a clicking sound, "Tsk, tsk, tsk." Almost unnoticeably, he shook his head back and forth.

The plant manager and another man dressed in slacks and an open-collar shirt came out to examine the load. They huddled together with the USDA area supervisor. I overheard them say something about more loads just like this.

My sample bag was taken inside, and as we cleaned up, I saw the woman from the USDA lab who recorded the results come rushing out, clutching a paper in her hands. The group of men read it. One rubbed the back of his neck, the others shook their heads. One volunteered, "I'll go make the call to the bank." The old farmer never saw the results. He didn't need to. The receiving docks were blanketed in silence.

Later I learned that the entire harvest from this vineyard failed so badly it was sold as cattle feed for a fifty dollars a ton. Rumors spread about the old man, a good farmer who had overextended and bought too much land on leveraged money. His sons had convinced him that with enough property, they could build a thriving corporation and boost profits. Now the children had left, seeking day jobs and a salary. The old farmer watched ranch after ranch lost, given back to the bank, sold out of desperation. The raisins I had inspected were from the original "home place" that had been used as collateral. It too was now gone.

No one talked about the situation; the silence said much. I spent the rest of that autumn working very quietly, wondering who was to blame. Occasionally out in our fields, I found a single bunch of grapes still left on the vine that the pickers had missed. Some were withering and drying on the canes into raisins. I'd stop and treat myself to a welcome snack. These grapes had been left hanging so long their sugars had collected and mixed with a fruity tanginess. But they were bittersweet in my mind. Even surrounded by my fruitful vineyards, I relive the stillness of that one moment of inspection. A story I did not yet fully understand. Only after years of farming and the slow, reflective taste of work have I learned to recognize the difference between sympathy for the old farmer and empathy.

How to Eat
a Peach

FIRST PEACHES

Weather report — Early July. Soon we harvest the Sun Crest peaches. Daytime highs are normal, in the high nineties, creeping above one hundred. Evenings cool — comfortable sleeping weather with your windows open. It's dry and hot. Visitors and out-of-towners hide inside air-conditioned rooms. I want to spend hours with my crops, sharing the heat with them. Though we have three peach varieties that start at the end of May and more varieties that take us into August, the Sun Crest variety is my favorite. They're the best-tasting and worth the wait. Summer doesn't officially arrive until I get my first bite of Sun Crests. Dad says that "the crop is made" and all I have to do is to bring it home. I wonder if he's giving me fatherly advice: "Don't screw up." Good peach-growing weather.

How to eat a peach begins with first finding the perfect peach. After thirty years I pretty much know where my prized Sun Crest peaches will first ripen. Usually it's along the eastern edge of the field where the morning sunlight initially strikes as it peaks over the

Sierras. There's a bit more sand in the soil there, so tree roots can take up water and nutrients easier. I have to keep an eye on these trees, because the ground dries out fast. A week ago, one branch of fruit surged ahead of the others and her fruits became the object of my pursuit of perfection. Somewhere in this cluster of fruits dangling in the morning light, growing fat with juice, the harvest will begin. "These first ones," I whisper, "are for us."

Each day, I pass by this spot a dozen times on my way to work in other parts of the farm or going back and forth from my home to my folks, about a quarter of a mile away. I stop to cradle the blushing gems, estimating how many days until climax. But I am not alone with my reconnaissance. I turn over one of the pristine, ripening fruits and to my horror discover a bird has also eyed these prizes. The creature too had been monitoring the harvest but lost patience and attacked the one side of a peach that was ripe, ignoring the other green half. A deep, jagged gouge has been carved into the flesh, a series of jabs has ripped meat from the interior. A hole about an inch wide and deep violated my treasure.

I want to rip the blemish from the tree and begin devising strategies to protect my firstborn – a mesh net draped over the branches, plastic owls strategically placed on a limbs, or in a moment of rage, a gun to scare trespassers off my land.

"Birds warn you when you're close to harvest," a Mexican neighbor, Gene Redondo, told me long ago when I was just a boy. "They can fly the entire field and will hunt for the ripe ones." I still remember his old, gnarled hands, stiff from age and years in the fields. He held up his right hand, shaking with a slight twitch. He leaned over and patted me on my shoulder, whispering, "And they return to the same peach, wasting nothing."

I must have been about seven or eight years old and didn't have a clue about what he was talking about. I just wanted to ride my bike and bounce over the irrigation furrows in the orchards and pretend I was a bird, soaring up over the treetops, escaping from this boring farm and melodramatic old men. But he was a nice man and I told myself I'd try to remember what he said, whatever it meant.

I release the damaged fruit, allowing it to swing back and forth in place. This will test if my old Mexican neighbor's tale has any merit. A few days later, I drive past the orchard with my truck window open. The aroma jolts me like a slap across my cheek. I sit up and coast to a stop. In the evening air, when the heat of summer rises and the coolness of nightfall begins to claim the surface, a fragrance of ripening Sun Crest peaches lingers near the earth. The aroma, like a siren, beckons attention. It will not be long.

Selecting the first perfect peach of the year becomes a quest. In the early-morning light, I return to the limb I had singled out a week earlier. Much of the fragrance has dissipated in the crisp air but a scent still loiters. I watch the sunrise reflected on the peaches. In the predawn shadows the fruit appears muted and dusty, a dull pink and green, darker because of the lack of light. I search out my secret branch, disappointed because the fruit doesn't look ripe, disheartened as I gently cradle one – it's still hard, firm, and without give.

But over my shoulder a slice of the sun dances through a gap in the Sierras. The peaches come alive. The colors shift. They blush a pink, a rose, then a fiery scarlet. The earth rotates and as the sun lifts itself, the distillation of light races through the orchard. A single leaf lies against the flesh of a peach. I brush it back and its outline is emblazoned on the surface, a shade of pale yellow-green, in sharp contrast to the surrounding reds branded by exposure to the light. I am reminded of *saggar* pottery; during firing, the flames of the kiln ignite a leaf or stick or other flammable material leaning against the vessel and the fiery image is burned forever into the clay.

I turn with my back to the sun. The rays strike my neck and back, a warmth generated by the massage of light. The colors are illuminated differently now; the ripest peaches seem to glow, not draped in reds but with a background amber that intensifies with each minute. I tilt my head and the angle changes. Perhaps it's the fuzz on the peaches. Each fiber embraces the light, and like a prism, certain colors refract and intensify. Like sunlight striking a haystack, the individual straws capturing then reflecting the elements of the moment – the early-morning or setting sun, the dampness of dawn dew, the glaring noontime sun, the vibrations of summer heat moving in waves.

Mornings – the first light passes through layers and layers of atmosphere. No different from in the evening, a setting sun blazing through what were white or gray clouds on the horizon a few hours before and now painted with brilliant lavenders and mauves, scarlets and oranges. But in this first light of day, I hold those same colors in my hands, a moment captured in these peaches, a sunrise fixed in the fruit.

The sun continues to rise above the massive mountain chain to the east. I glance upward and have to squint; already the illumination has changed. I've been trained to see objects like a peach or leaf or branch but not light. My eyes dart from fruit to fruit; slivers of light dance across the red blush. I stretch for the one that seems to shine. But it's just out of my reach. I'll have to drag a ladder to the spot and work for my treat. It's only proper, though; this is not a convenience food. Rewards wait for those who are willing to search.

The peach that I've watched for weeks has a slight give – I can gently apply pressure to the surface and my fingertips leave a very mild indentation. I can pick it now. I sample my first peach with a single bite from the pointy end of the fruit. The peaches are ripe. I like to graze with the sunrise, when the fruit still carries the night chill, not cold like out of a refrigerator but like a cool morning. Temperature seems to add to the flavor, increasing both the acidity and the sugars and transforming them into a tasteful balance. Parting some leaves and branches, I discover four more peaches on a lone limb that extends high above the canopy. I have to climb down and reposition my ladder, imagining the fruit to be gone when I return moments later, as if they were a mirage, a dream sequence from a Kurosawa film. They're still there and I pick them all, using my overturned straw hat as a makeshift basket, bringing them home to family.

In the center of our kitchen, we hover around a butcher-block island. I carefully slice the first, avoiding the pit and revealing a creamy yellow flesh, juice shimmering in the light. Lips pucker and taste buds dance. We can't wait.

In our annual summer family ritual, usually Korio begs for the first bite and inhales the flesh, ready for more before I've handed out the rest. The ladies of the house are next — Marcy and Niki each get a slice. Nikiko's old enough to distinguish between good peaches and great ones. We've challenged her to make that distinction and say, "It's okay to ask for great." She's taken it to heart and at times is hard on herself when and if she feels underachieving.

Both lick their lips in preparation. Each slips a slice between lips and savors the juices spreading over tongue. Then they bite and the explosion of flavor bathes the senses. They smile. And nod. All is well. I feel as if I've passed the most important test of the year.

LOSING VIRGINITY

Weather Report — Hot. Recent night temperatures never drop below seventy degrees. At eleven o'clock the night before, it was still in the high eighties. Forecast today: over one hundred with tropical humidity. Not common for us — we're used to "dry heat." Already at 6:00 a.m. I'm sticky, just from walking the fields. The humid air mixes with the dust from weeks without rain. Daily the air grows ugly and brown — from cars, cities, industry — from people. By 7:00 a.m. I can hear myself breathing harder.

But the first sweat of a workday feels clean. Sensual. Natural. Part of a rite of harvest. Mornings in early July, the sun rises at 5:30 a.m. and we can start

picking if the peaches are ripe. Their red blush is like a beacon, dark and mys-
terious in low light, gradually glowing with first sunlight. The trees hang heavy,
branches lean toward the earth loaded with peaches. Red, voluptuous peaches.
Ripe. Juicy. The heat and humidity bring on the harvest. Every morning is calm.
Quiet. Peaceful. "Full of potential," I tell myself. The sun rises with hope.

When did I taste my first real peach? As a child, I remember the juices running down my face and flavors exploding in my mouth. The sweet nectar danced across my tongue as I sucked the meat down my throat, gorging myself while smacking the lips. Until I was a teenager, I was fat, and all my white T-shirts had the same pinkish-red stain running down the front of them and arched toward the edge of a robust tummy.

After drying it on the clothesline, my Mom would hold up my stained shirt in the sunlight and say, "Peach juice," while shaking her head. She wasn't angry, because she knew it was a good sign. Kids enjoying their family's fruits seemed like a good thing.

What happens if you don't have memory of real peaches or fruit? Lacking experience, you eat quickly. First, you limit yourself by only allowing your eyes to guide you, searching for something red to bite. In your hand could be a peach or plum or baseball; you don't feel anything except something hard. Next, with a rapid, unthinking motion, you bite as if it were just a reflex connected to "food-in-hand." (Unfortunately your sense of hearing may also be stimulated. Apples crunch, but I don't think peaches should.) Finally, as if to dispose quickly of something that lacks flavor, you hardly chew before swallowing. Then you start the next bite before finishing the last. No need for napkins or shirtsleeves to wipe your face with; no time when you're racing.

But eating is a physical act demanding dexterity of the senses. I start with my fingers, spinning the fruit in my hand, instinctively searching for the soft, occasionally gushy side of the fruit. I caress the skin, feeling the shape and bulges, stroking the suture, a gentle seam, a natural cleavage.

I use the fingertip pressure approach, softly massaging the surface, a squeeze to detect "give." Some think of "give" as resistance to the pressure of your hand. I find that perspective too one-sided, as if you squeeze the flesh to see how much pressure it can withstand until it breaks down and "gives in" to your demands.

Others use an open-palm pressure technique, setting the fruit flat on the hand, then wrapping the hand around it by moving from the palm area first, then the base of the fingers next, never using the fingertips. The meaty part of the fingers cradles the fruit, a hug of the flesh. A sign of "give" is when the two surfaces — your fleshy fingers and the meat of a peach — each press against the other.

I don't use palm pressure, but I often hear nutritionists refer to this method. (I learned this technique from a training program in conjunction with various agricultural trade associations trying to devise a nondamaging method for consumers to squeeze their produce. Produce managers hate seeing their shelves full of dented and squeezed-to-death fruits, often left in front for the next buyer's eyes.) Instead, I feel with my fingers while rotating the fruit; I can cover the sphere within a few seconds. I dance across the surface, skimming over the curves, the pads of my fingers much more sensitive than the tips. I don't hunt for "give," it finds me. I'll detect a softer feel along one side or in a general area — a cushion against my skin. It's as if the fruit pushes back.

With some fruit I can immediately notice the "give," and I'll quickly stop and put down the treasure, knowing it is ready. Most often I'm not sure, delaying any judgment as I sift through my assessment to what is there. The tip of a peach usually ripens first, while around the stem the flesh is hidden from sunlight and remains firm. I don't know if it's the sugars that soften the fruit as they spread unevenly throughout a peach, or if it's the basking in the warmth of sunlight that stimulates cell growth, absorption of moisture, softness and sweetness.

Next I close my eyes, paying attention to the aroma. Only inches from my open mouth, I detect the scent of a peach. It's almost impossible to taste without smelling first. Researchers claim it's the loss of smell when we have a cold that accounts for food having no taste.

I slowly lick my lips, force myself to take a deep breath, carry the bouquet deep into my lungs. The perfume spins my memory like a passing fragrance in a crowd of people, and I look up, searching, trying to locate an image, an old girlfriend, an experience stimulated by an essence. Sight, feel, and smell, they all then contribute to taste.

My mouth waters in anticipation. I find myself almost drooling – years of enchanting flavors have trained me well. I feel my teeth sink into the succulent flesh, and juice breaks into my mouth as I seal my lips on the skin and suck the meat. I roll the first bite in my mouth, coating my tongue, bathing it in the nectar and fibers. My taste buds pucker slightly, excited with the natural blend of sugars and acids. I enjoy yellow-fleshed peaches the most. Unlike white flesh, yellow flesh blends a mild level of acid into the mix of sugars, resulting in a sweetness not like candy but with a certain tanginess that stimulates.

My lips smack and the sound of eating joins the sensory feast. A kiss of air mixes with the flavors as I chew with my mouth open. Like a good wine breathing, the taste of the peach is getting better with each moment.

I eat slowly and sigh, thinking of the times I've wolfed down my food, shoveling it from plate to mouth, contents funneled into a cavernous throat to fill space inside. Now, even after a single bite my mind wanders, flavors drenched with memories. A rush of pleasure. Senses excited. Taste lingers in your mouth. For a few moments, I believe I'm young. I smile, then open my eyes. I think of losing my peach virginity and the sensual pleasures of eating. Each time it feels wonderful.

My nine-year-old son, Korio, is already a veteran of peach harvests. He's small for his age, a tiny guy under four feet and a scant fifty pounds. When he sinks his small teeth into the flesh, most of his face is hidden behind the huge peach he has chosen, lifted with both hands to his child's mouth. His bite barely dents the monster-sized fruit, since he loves selecting the biggest and ripest. And why not? It's one of the rewards for being a harvest son.

Korio doesn't bother to lean over as he munches; the juice and flesh drip down his face. He doesn't mind, yet I've seen adults physically grow uncomfortable when they see a dribble collect on his chin, drips almost ready to fall, crying out to be wiped. The grown-ups fight the urge to clean his face. Instead, his shirts are stained, as mine used to be, initiated by the rites of a proper peach harvest. Sometimes his small hands can't hold the peach, and it drops to the ground and rolls in the fine loam dust of the barnyard. Unfazed, he simply picks up the fallen fruit, makes a feeble attempt to wipe it off, and with his next bite the earth adds texture to his treat.

When he lowers the peach and grins, the amber juice and peach meat paint a clown's smile across his face. Before we lost our old farm dog Jake, the aging golden retriever would occasionally pull himself up and plod over to my son. As Korio squatted, slowly enjoying his summer fruit, consumed by the moment, Jake would help clean up that sticky face with long, even strokes, licking my son's cheeks. Korio closed his eyes tightly and squeezed his face together until the tongue bath stopped, then returned to his harvest meal.

WINDOW ART

Weather reports sometime give humidity in percentages. We start mornings in the high teens, rising to 60 percent by evening. (East Coast much higher. I've felt it only while visiting but never working. It must be dreadful.) Humidity physically draining. Tired just by moving. Weak.

Humid day predicted. At first, daybreak feels rather pleasant, then by mid-morning lungs start to suck air. Breathing becomes labored. Workers begin to "dog it," and I don't blame them. Fields like saunas, dirty saunas that last for hours. Work clothes soaked with sweat by noon, not just shirts but pants too.

Once in this type of weather, I had to set the water pressure in fifty irrigation valves, opening each brass gate inside the cement risers, then digging weeds around each stand and shoveling dirt to make sure the water ran down a furrow. It was midafternoon by the time I got to the job that had to be completed by evening when I'd start the deep well pump.

By the fourth valve, my shirt was dripping. Salty sweat stung my eyes. I grew nauseous, challenged and pushed. When I finished, my pants were so wet I left a soggy imprint when I leaned on things. My leather belt was stained with dark perspiration. Wet pants — an indicator of the hottest and most humid days of the year.

Can't get enough liquids. I offer my work crews some sports drinks with electrolytes. Try to explain it will help them feel better. I don't know the word for "cramps" in Spanish but I told the men about my nightly chingazos, *fights, with my body, legs tightening and muscles cramping. That's why the terrible-tasting drink will be good for them. They try a glass and do feel better. They want more.*

Work slows with the heat and moisture. The trees shut down, stop producing sugars as they shift into a survival mode. I worry about the hanging peaches. Will the trees abort them? Or will the fruit simply stay small and go soft?

They sit on the kitchen windowsill. Waiting. Hoping for perfection.

Ripe fruit when picked will continue to ripen. You can smell it — they have a natural peach perfume called ethylene. Other fruits produce a similar scent — apples and bananas, to name two — but the scent of peaches seems more delicate and fragile. While the sugars of a harvested peach remain the same even after the peach is picked, a mature fruit continues to change the ratio of sugars to acid. Researchers call this "internal ethylene," and it's an indicator of ripeness. Fruit when picked green will remain green. I swear instead of producing an aerating fragrance, green fruit does the opposite and absorbs the aroma of its surroundings. It begins to smell like cardboard.

With my first ripe peaches, I've tasted perfection, but now how do I share it? The real challenge lies before me – how to pick a peach for the market, knowing it will take a day to pack and deliver to cold storage, where it may then rest for another day before shipping. West Coast delivery in a day, Midwest in three, East Coast in four or five. Then at the retail level, a day in their "walk-ins," cold rooms, before displaying. Can perfection be delayed? Taste and timing abruptly married to each other.

I have friends who sell at farmers' markets. They can pick riper fruit with shorter lead times, but most have found that not all consumers want their fruit juicy ripe on that day of purchase, and if they can't sell their soft fruit in a day, it's lost. They too struggle with the challenge of suspended excellence.

So every year I pick sample peaches, set them on the kitchen windowsill, and wait. With a felt pen I mark the date that I picked them. The marker bleeds into the fuzzy skin, but it's legible. I circle a yellow area and write "g-r-e-e-n-?" I circle a softer spot to monitor. Initially I wonder. Too hard? Too ripe? Every time I pass through the kitchen or come in from the fields to check messages and steal a quick drink of ice water, I examine the peaches. Later, I study weather reports and monitor temperatures and humidity. Time to pick a tree? An acre? A field? We pick fruit in rounds, taking perhaps a fourth to a fifth of the total fruit with each pass. My windowsill should help guide my scheduling – but will the picked peaches be ready two days from now? Or do they need three? A test of ripening. When I start harvesting, I become a connoisseur of worry.

With peaches, I can't see perfection. Other than taking a bite out of each one, I'm not sure how to measure it. The perfection of our first "family peaches" may work for the short hundred-yard trip from the field to our kitchen, but what about the restaurant in Berkeley or the produce section in a Minneapolis co-op? Staring at the collection of peaches sitting on the kitchen windowsill, I have time to ponder this. I carry the image out with me while I do monotonous tractor work, turning the question of perfection over and over in my mind, spending way too much time thinking about it.

I daydream about peaches, clenching my teeth at the challenge before me until my jaws ache at night. I've developed a slight trigger finger in my right hand from unconsciously opening and closing my fist while I desperately dream of achieving a consistent standard of excellence. I've imagined my task like a Greek odyssey — a quest to please the gods and their whims and jealousies as they toy with us mortals. Aphrodite must have been a lover of peaches, their scent and aroma, the purity of the blush, the flavor of the flesh. Peaches as an aphrodisiac — until my ode turns into a classic tragedy and my hopes are crushed when I realize peaches weren't introduced to Western civilization until after the travels of Marco Polo. My furor reaches a biblical crescendo as I imagine peaches closer to sacramental wine than any other fruit. I believe they should have served peaches at the Last Supper instead of bread and wine. Peaches are alive; they grow and change, the pulpy meat aging with a sweetness of life, the flavor subtle. They are grown by stewards of the land, hardworking, toiling with a sense of respect and awe. Fresh peaches — not something baked or fermented. A huge fruit cut into thirteen — no, twelve thin slices, eaten slowly so that the spirit of the land and the farmer and the won-der of God's nature is tasted in its full glory. Peaches, the fruit of heaven? My vision is suddenly clouded, though: I believe Eve and her apple gave all tree fruits a bad reputation.

Later in my daydreaming, I start asking questions with no answers.

A fruit farmer's *koan:* what's the sound of a ripe peach falling in an empty orchard? But my orchards are filled with weeds and cover crops, while spiders, small mites, other critters thrive in the undergrowth. Answer – it's never empty. I picture my peaches in the hands of a chubby Buddhist monk, who grins with each bite. Perfection is not about precision. The perfect peach is not a processed product, it's not about efficiency, it's not controlled. Once I toured an organic baby food processing plant. They had purchased tons of my peaches, which were destined for the hungry mouths of infants. The quality control manager and I watched my peaches tumble from a conveyor belt where they had been sorted, the damaged fruit tossed aside. Then the fruit went into a hot wash bath and disappeared into a large metal box, where they were blanched, the skins and pits were removed, and they were pureed. I couldn't see how it was done. Instead, from the other end of the contraption came a stream of liquefied peach, ready to be frozen and stored until mixed and combined with water and a few other ingredients to make a peach baby food.

The quality control manager bragged, "Bring in any peach and I'll fiddle with the acids and sugars. By blending fruits, I can fix any peach," he said. That wasn't the perfection I sought.

The art of growing peaches remains a wonderfully natural process with an expectation of differences. No two peaches will look, feel, smell, or taste the same. Then is perfection achieved when things are out of control? I've enjoyed some of my best peaches in years when chaos seemed to reign, when weather was always variable, field conditions constantly fluctuating, and I never felt the same from day to day. Why should I expect consistency from peaches?

The peaches on the windowsill change. I monitor them two, three times a day. Tones shift from light yellow to darker gold. Flesh softens and aroma grows. The areas I had circled with a felt pen were yellow but soon blush with ripeness and blend with the rest of the hues. I slice one and the flesh glistens in the light, juice inviting . . . no, daring me to bite into it.

I hesitate just before sampling, doubting the peach has grown sweeter, worrying that the best I could do is to ship a very good peach to market. "Good enough" was a statistical term from an Ag Econ class I had taken, the gray area of a graph where consumers' lowest expectation levels are met. "Good enough" delivers average market prices. "Good enough" would not create disappointment nor anger. I could easily hide behind "good enough." As the slice nears my mouth, I immediately smell the aroma. I slip the sample between my lips, and it melts in my mouth. It's wonderful.

"GREEN!" had been boldly written on the fuzz of one peach. But what exactly did I mean? I discover a patch of yellow. I slice the fruit, eat another slice, and can detect a slightly crunchier portion with a bit more acid, firm, yet the surrounding meat satisfies. I wonder if I had not been searching for the greener flavor, would I have even noticed?

Every year I need to prove to myself the simple yet still magical rule for perfection: ripe fruit continues to ripen. The concept boggles me for some reason. I remember a summer in the early 1980s. Midseason peaches in oversupply, Georgia and South Carolina with bumper crops, flooding the East Coast markets. Few California peaches ventured east of the Mississippi. We all suffered terrible prices. I was angry and broken, wanting to get our damn peaches out of my sight. If I picked ripe – ten cents a pound to me. If I picked green – ten cents a pound to me. That summer I worked hard just to cover harvest expenses and offset a portion of the production costs of pruning, thinning, fertilizing, controlling pests, and irrigating. My year's labor was donated.

I remember saying, "So why pick right?" and I didn't. I learned a lot about green peaches – they stayed green. Instead of going soft in cold storage, they shriveled and dehydrated; less water content meant that even the tart fruit seemed better because the ratio of moisture to acid changed – less water increased the percentage of remaining sugars. They didn't ripen sweeter, they just didn't taste as terrible.

Green peaches kept for weeks. No bruising from rough pickers. Hard as rocks, a lethal weapon if you threw one at someone. No damaged flesh meant they all arrived looking quite well. No adjustments in price. (But why would a retailer want to bicker? At fifteen cents a pound, they had a lot of margin to play with.)

The experience left scars – I had sent produce to market that I would never feed my family. It was right for all the economic reasons but terribly wrong. A broker laughed at my worries, welcoming me to the world of business. A young farmer learns a harsh lesson. I had lost my naiveté and didn't like the sensation, even though the salesman was right and I knew it. I did learn that at times you had to accept market conditions and move on, because the most important thing was to survive so that you could have another harvest. But I hoped that summer would never be repeated.

So every summer after that I have worried about being forced to pick "green" and having to face the awful peaches of that summer. I have avoided a repeat of that depressing season, but continue to struggle with a fear of failure. Will that green spot on a peach sitting on the window sill actually disappear? The bitter fruits of the past linger, a ghost of self-doubt haunts each harvest.

I'm stupid to worry so much. Will I always lack confidence in a peach? Even though I was born and raised on a farm, as a child I had often ignored the world around me, wandering to imaginary places far away. I had memories of those days that I cherished and frequently returned to, yet it wasn't until I came back home as an adult that I realized the complexity of growing a peach. I had to learn my own style. All farmers, in their own way, are self-taught. We work in isolation like the artist trying to find voice or the craftsperson seeking mastery. Or in my case, searching for the peach with the perfect taste.

So after decades on the farm, what have I learned? Perfection becomes a product of synergy. I begin to make an ongoing list of what makes the ideal peach. Variety — in my case I'll work my entire life with a few, including the one called Sun Crest. Old-growth, seasoned wood over decades and preferably generations. Soil that matches. Water with care. Fertilize naturally. Paying attention to what works. A caring passion.

I suspect the more I search for excellence with an increasing challenge at harvest, the sweeter the peach — a direct relationship to the amount of effort required to claim a treasure. No different from what I witnessed at one of the best produce markets in the San Francisco Bay area, where shoppers have to search through a large pallet of individual boxes of peaches to select the ideal fruit. Perhaps when individuals make their own choices — as opposed to being presented with the pristine stack of fruit where each and every piece more or less looks exactly the same — they value the end prize more and find it tastes sweeter?

I call my peaches basking in the light of our summer mornings "sill life at harvest." Sitting in the window allows for close observation of natural details; the peaches become my personal focus group. I live with them for weeks. I'll find myself talking to them and realize just how isolated farming can get. When my nine-year-old son catches me in a deep conversation with a blushing peach, I whirl and he just grins. He understands precisely, because he's nine and knows how to talk with imaginary friends.

Within a week, a row of them line the sill, all at various stages of ripeness from being picked on different days. I'll usually start a quasi-scientific experiment, placing a few in the refrigerator at temperatures of forty-five to fifty degrees, possibly the worst environment for them. The chill isn't cold enough to slow the ripening (our cold-storage facilities hover in the mid-thirties but not quite freezing). But it's certainly not warm like the outdoors, where the natural evolution would continue.

Studies have shown peaches become mealy when stored at those temperatures, a combination of slightly dehydrated (which is what refrigerators are supposed to do — the better ones try to remove the humidity) and uneven maturing from the cool but not cold locale. I note my observations of peach behavior when mixed with the emotion of the moment — sometimes quite confident in their quality, the next moment terrified that the truck carrying the load to Boston is just like our refrigerator and the peaches will arrive shriveled and with their meat coarse.

Then I'll stop the test, my imagination running wild and confidence crumbling. I'm too exposed. Once a produce buyer phoned me during one of my anxiety attacks. He was a good businessman and quickly detected an opportunity to get a low price for my peaches. He let me talk, and during my monologue, my price kept dropping with each lament. Insecurity translated into lost dollars. Vulnerable artists shouldn't be selling their art.

Midway through the harvest, I'll revive my study of nature by lining up a string of peaches all picked the same day, green fruit on the left with riper fruit on the right, a typology of ripeness. Would the green ripen? Could the ripe get even riper? Second- and third-round peaches — those from the second and third times we enter the fields to harvest — seem the best quality. I can pick those fruits a little greener, and after a few days, they're fine. Even ones almost too soft at harvest will hold without bruising or the interior flesh browning. My sill life tells me so.

Everyday peaches from a typical harvest, subjects from nature in their natural light. Perfect doesn't mean the skin will be completely even and glasslike smooth. Some peaches have a more pronounced tip or a bulging suture. What does it mean? I don't know. But their shading attracts my attention and curiosity. The unevenness adds something. Can you taste the texture? I know I can't separate what I see from the taste as I bite. It is part of the flavor.

A few summers ago, I found a mutant branch on one of the peach trees with a blossom on it. A new blossom while we were harvesting in the middle of July. I brought it home, and Marcy set it in a small vase next to the windowsill peaches. We added some sprigs of leaves, then a small woody branch dark with twists and coarse bark. I placed our family fruit label next to these completing my painting. The objects stayed for days until the petals fell one by one.

Later that summer, I attended a writers' conference, and during my reading, I asked the audience, "How do you know if a peach is ripe?" I envisioned my sill life and wanted to share my passion about "visual taste."

But my delivery was interrupted. A young male voice yelled out, "By the color of the peach on the can," and was answered with a roar. The sill life of pop art.

Is affordability part of perfection? Organic produce with its often more expensive prices has been called elitist by some. But it's not my choice. In the world of fresh produce, it's still the Wild West with market forces determining my prices. Lots of peaches, lower prices; scarce fruit, high prices. Most of my fruits are sold wholesale; someone else sets the price for consumers at the retail level. Almost every summer, either peaches, plums, or nectarines die in the marketplace from oversupply; boxes are "dumped" into the supply, and prices bottoming out at a nickel or dime per pound for the farmer are quoted. But in the grocery, rollback prices will either be delayed or will never happen. In 1987, my Sun Crest peaches got five cents a pound. Meanwhile the stores sold them for a dollar a pound until they could go "on ad," and then they dropped to sixty-odd cents a pound. Affordability had little to do with me.

Can I rationalize a dollar or two a pound for my fruits? Yes, by watching a child eat a real good peach. The fruit tastes so good, and the child eats it all, slowly, with a patience we rarely see in a younger generation. I wonder how many cheaper and awful-tasting fruit are tossed away or sit in a basket uneaten until they shrivel or decay and are tossed.

Candy is expensive, as are breakfast cereals and snack foods. Twinkies cost $3.50 to $4 per pound, depending if you buy them in twin packs or a ten-count box. We normally don't buy ready-to-eat foods by the pound; otherwise people would know that pizza can easily cost $5 a pound, a quarter-pound burger, after it's cooked, about $4 a pound.

What is the proper price of something good to eat? At times during harvest, I'll phone my brokers to find out the price of my fruit and occasionally realize that I grow peaches most of my friends can't afford. But that's not entirely true. Most of my friends have money; they just may not want to pay $2 a pound for peaches. They're not used to it because most of our food is cheap. The average American spends under 11 percent of income on food, one of the lowest percentages in the world.

While following my peaches to market as they ripen on the windowsill, I don't worry about prices. I'll keep track of the market and pencil out rough estimates of expenses and income. Farmers plan for the worst-case scenario when prices crash so low that harvest expenses exceed income. We've learned, painfully, how to stop picking and leave fruit hanging on trees. This has happened only a few times in my life, but there have been many other times when we still lost money and I kept picking in order to earn back some of our production costs of pruning, thinning, and fertilizing.

I rarely hear from brokers and salespeople unless there's a problem. So while I long for some response, I'm also terrified about calls: from the business perspective, no news implies good news. The farmer side of me, though, wouldn't mind a few kind words, like a soft whisper of encouragement.

But one of the worst moments in my harvest had nothing to do with prices or feedback from consumers. One morning I monitored the peaches lining our kitchen windowsill and discovered some of them were gone. Like teeth missing on a toddler, harvest gaps stood in my collection of fruit. All the pieces I had carefully dated, with extra markings indicating a green area or unique characteristic I was studying. Now some had disappeared.

Marcy then informed me what had happened. For dinner the night before, we had had a delightful summer fruit salad. It was wonderful-tasting, but it had cost me a season of sill life study and research. It was worth it.

My *baachan*, my grandmother, taught me how to eat a peach. She'd sit on a small wooden stool and, leaning over a metal bucket, peel the fruit with a paring knife, a steady ribbon of skin dangling from her skilled fingers as she rotated the fruit in her old, callused hands.

Occasionally she'd stop and like an innocent child, she'd steal a slice from the golden flesh, and quickly sneak a piece into her mouth. I watched her close her eyes and they seemed to tremble, muscles of an eighty-year-old involuntarily twitching and dancing as if lost in a dream.

Baachan savored the flavor. A satisfying glow gently spread across her face. Not a smile or even a grin, just the look of comfort, relaxed, soothing, content. I thought of that image even after she died, want-ing to believe that would be the look on her face forever.

Each time I return to that moment, I can see her hands alive and dancing, massaging the skin, spinning the peach to find a good spot to start. Her mouth rests slightly open and I can then hear her smacking of lips together, a soft, wet sound as she mouthed an imaginary bite in anticipation. A peach aroma surrounds her like a cloud of perfume.

I remember the taste of those peaches as something special, but I imagine the flavor was different for Baachan. I don't think she tasted the peaches in her hands. Instead, a rush of memories and harvests from her past filled her with each bite. These were no longer my dad's peaches or even from our family's orchard. The ideal fruit quickly dissolved into her own thoughts.

Peaches sitting on a kitchen windowsill like a shelf of library books, each with its own story, and with luck, the perfect peach disappears with each bite and becomes nothing but memories. I don't offer a "money-back guarantee" because perfection doesn't work that way. I enjoy the creative aspects of the art of farming. Once a season begins, wonderful changes unfold, often unpredictable. My goal then is to help create a perfect memory.

Over a decade ago, Baachan and I shared that perfect moment. I've spent years trying to reenact that scene. Closing my eyes. Smacking my lips. I smile and gradually I too lose myself in the flavor of a perfect peach.

Personalized
Produce

Entries from my farm journal during one harvest day when things are not working:

— Trailer loses a bolt, one of the two that holds the hitch and the frame together. It looks like a wounded bird, flopping along, dragging one side, injured. Panic to fix it within an hour but first we have to limp home from the field.

— Tractors run rough, sputter, and die. Air filters in bad shape. I pull one out, can't remember changing filters in years. Dust is caked on the surface. As I tap it against the cement, a small pile collects under the membrane. The engine breathes in this stuff? It coughs as I urge it along.

— Truck temperature gauge stops working on hottest day of year. "Don't need one," I rationalize. I then drive in fear of overheating, adding another level of stress.

Is this all my fault? I squeeze in the time to fix the trailer, make a note to change tractor air filters, and will tackle the truck temperature gauge in the off-season. And tomorrow something else will break, adding to my expenses and costs. Yet despite all this I can still have the perfect harvest because perfection isn't about making money. It's personal.

I took this to heart when I developed my PLU – Price Look Up – sticker. As computers found their way into grocery stores, checkout scanners have become the standard of how business is done. More and more products are required to have some type of electronic identification – most often used is a "sku," an electronically coded stripped bar on packaging that charges the correct amount and provides categorized inventory control and instant sales data. Until recently produce was still identified visually by the checkout clerk, whose hand entered a generic code and whose eyes looked up the price on a handy chart as the product was weighed. It was slow and especially confusing with the explosion of specialty fruits and vegetables and workers who couldn't tell the difference between a nectarine and a peach, let alone distinguish the yellow-flesh varieties from the white-flesh.

Peaches don't have skus, probably because the long labels can't fit on a piece of fruit. But we use PLUs, oval stickers about three-quarters of an inch wide by half an inch high. Some fruits have been stickered for a while. Bananas and apples already had logo labels decades ago, so adding a number code was simple. Many other fruits have smooth surfaces and the food-grade glues can easily adhere. But peaches have fuzz, especially my older, heirloom varieties. Keeping on a tiny, smaller-than-a-postage-stamp sticker was going to be a challenge. A neighbor, Dave, who helps me pack my fruit and is a closet engineer, designed an electronic reel and triggering device that sits over the continuous belt of peaches. He initially grew frustrated when more labels ended up on the cement floor than on the peaches, but fine tuning with a constant monitoring by his wife, Dorothy, or three children helped the machine gently kiss most of the peaches with my PLU stickers.

It's almost impossible to sticker each peach. During my peak harvest week, I'll pick and pack a quarter million peaches. Labeling each one isn't where I think the priority should be. Besides, rarely do consumers buy a single peach at a time, so retailers more or less accept a 60–80 percent level of stickered fruit, which is very high for my fuzzy friends.

I had initially balked at the idea. Perfect peaches marred by a sticker? How could tree-ripe peaches tolerate a machine that jabbed the surface and punched each fruit? Wouldn't it leave a bruise? Some retailers leveraged pressure: "No sticker, no sale." So I conceded and try to label most of my peaches, but only after I saw Dave's handy little machine, with his family's help, rubbing them the right way.

But the biggest question is what to include on the sticker itself. The industry code requires that for "organic, large and yellow flesh peaches grown west of the Mississippi," farmers have to use #94038. Nonorganic or small or white-flesh ones use a different number, East Coast peaches another code. That's all that is required. Soon, savvy marketers realized you could add a company name, a slogan, or place name promoting where something was grown. I attended a marketing workshop and the facilitators announced that the next evolution of the PLU sticker would be to add your web page or e-mail address.

Ironically, the technological evolution of the retail produce industry has changed how my fruit can be sold. In the past, produce was rarely personalized. Until the 1990s, we never had our name associated with our peaches at the consumer level. For forty years, only two or three people had ever heard of a Masumoto peach, usually our broker and a few select buyers. In the 1980s we worked with a large packing shed where our fruits were pooled with other farmers'; then even our own sellers didn't know our peaches.

During those years that we worked with the large packing shed, I remember once trying to find out how my fruits arrived at market. I had to call a sales office fifty miles away, and after being transferred from broker to broker, I talked with a lowly accountant (who actually was friendly – I imagined he didn't get to interact with many human voices during the course of his workday). He couldn't tell me much except by inferring from the prices our fruits received and how much was still in cold storage that someone didn't appreciate the peaches. My peaches consistently received a dollar less than market prices and the inventory listed a couple hundred boxes of week-old fruit ready to be "dumped" at any price. The manifest seemed to indicate those were my peaches, but he'd need to do a visual confirmation. I appreciated his honesty, and the word "dumped" hurt me (a term the sales staff would never have used when talking directly with the farmer). None of the brokers had mentioned the pooled fruits were in trouble – they used terms like "slow movement," but I was convinced they didn't know for certain whose couple of hundred boxes of peaches were still sitting around. I knew then they didn't care.

Initially, adding my name to identify my fruits scared me. Farmers have grown accustomed to working in obscurity, guarding their privacy, thriving in anonymity. We often feel uncomfortable about exposing our own fields to the public, sensitive to responses and reactions. A neighbor driving by is fine, a neighbor who understands the meaning of thick patches of weeds that got away or this year's struggle with some pest or disease. Revealing our private sides to the public seems frightening. I feel naked. Our reputations are at stake. Does the audience understand just how much of our world we have no control over?

Broadcasting my name also meant I would be the one who received complaints or read an e-mail describing in detail the shortcomings of my labor. Makers of other products like cars or branded food products encourage this sort of feedback, guaranteeing satisfaction, but how many times does the actual owner answer the phone? I doubt if members of the Ford family have to listen to an irate owner complaining about a car's design or poor paint job. Is there really a Mrs. Butterworth who patiently nods her head in empathy with a parent outraged when syrup sugars crystallize?

Adding a web page address to my sticker would invite people to read about our farm and contact us. The feedback worried me. After all, what could I do except wait till next year? I couldn't fix any of this year's peaches, and even if I did make major adjustments, I figure half of what makes a good peach is still out of my control. Yet perhaps the responses wouldn't be complaints but more like "reviews." I know many artists who claim not to read reviews but in private, secretly devour them. I could do the same and put aside negative comments until after the harvest was over and I was in a better frame of mind. Then, on a cheery day when life was going quite well, I could sneak a peak at the comments, trashing those overly critical ones, claiming, "They don't understand me." With other notes I'd adopt a defensive strategic approach, carefully reading the early part of a note, where usually the positive complements are sprinkled, looking for those delicious words like "great" or "wonderful" or even "nice" while searching for the "but" transition signifying the bad stuff to follow. I'd glance at the rest and if it got ugly, I'd literally put it away.

Farmer artists have delicate emotions and egos, like most other artists and craftspeople. I've learned that my sensitivity is where I draw my creativity from, yet it easily leads to overreaction. So I plan to be grateful to those who let me know of a problem with my peaches and also thank them for letting me not respond.

Finally, I convinced myself to add "Masumoto Family Farm" to the sticker. I was confident of my peaches' quality. I'd offer my best and stand behind my name. I thought of these as signature peaches. I still cringed, though, when I knew a few slightly green ones slipped into a box.

I received a note from a Japanese-American woman. Her father was living in San Francisco, a retired engineer and currently dying of cancer. He had grown up on a farm, and after reading some of my books, he and his daughter found my peaches at a store and sampled them, much to his delight. "They sent him back to his youth," she wrote. His latest treatment regimen was depressing him. His life might be prolonged for a few months but they both knew his time was short. She desperately asked if I'd come visit, just to talk.

We sat and had lunch. He told me of their family farm outside Sacramento. They grew everything from vegetables to plums, and of course some peaches. He had drawn a map of the home place, probing his memory of what was planted in each field, how they irrigated them, and the ones they had to "let go" when the prices grew depressed. We talked like old neighbors, lamenting the tough times for farming and the hard work invested. We joked at the crazy weather and the ways farmers patched up equipment or tried to trick water to go uphill. (Through a series of siphons and "checks" in a furrow, water can push up a slight incline.) We sat in silence and stared at the few photographs of the immigrant family, proudly standing in a field they called home, or in front of a "new" used tractor that seemed to shout out that they had become established in America.

Near the end of the visit, he thanked me for telling stories about farming. He remembered the hands required to work the land. He knew of the flavor of the perfect peach and what it meant. He cried. I sat in silence. Suddenly I realized it wasn't my peach he was reading about in the pages of my books. They were his peaches and the stories were no longer mine.

Food
Editors

A GROUP OF TEN food editors from *Better Homes and Gardens*, *Sunset*, *McCall's*, *Country Living*, *Good Housekeeping*, and other major national magazines came to visit my farm. They were part of a tree-fruit-industry-sponsored week of touring California orchards. Wined and dined, they saw modern farm operations with sophisticated technology and production methods. They walked through a shed that housed the latest in computer-assisted high-tech sorters and packing equipment and sat in seminars about the recent research programs geared toward producing a consistent piece of fruit. Industry leaders hoped these editors would then write articles promoting peaches, plums, and nectarines in their upcoming issues.

As a side excursion, the group had requested a visit to a small organic family farm. We began with an early-morning brunch in the peach orchard. I read stories about family and farming and conducted a farm tour by walking leisurely through the fields and explaining what I see and feel as a farmer. A few of the editors found it hard to slow their stride. "Farmers become connoisseurs of dust," I announced and explained the regional cuisines of soil types as they related to a farmer's palate.

I ended the tour with an interactive challenge. "You have written about how to select and cook with produce. But how do you pick a ripe peach? Is there an art to choosing the perfect peach?"

I briefly described the farmer's dilemma. If I waited until a peach was soft and gushy – the way I like to eat my peaches – the fruit would barely make it from the tree to our kitchen before it was bruised. It would be overripe. So how do you harvest knowing there are a few days between the farm and the grocery store? How do you anticipate taste and forecast flavor without sampling each peach?

As I spoke, Marcy handed out buckets labeled with the names of the food editors. The group was to pick peaches and I would store them in our refrigerator for a day or two, as if the fruit were in cold storage and later in a refrigerated truck speeding across our highways. Meanwhile, the editors could return to their East Coast offices and we'd then FedEx their own personal peach harvest. They could verify how well they knew their farmer "within" and learn a little about the gastronomic realities of farming.

The editors were game and up to the task. They wandered through the orchard, suddenly studying each fruit with a different agenda. Some wanted desperately to ask Marcy or me which ones we'd recommend, but stopped and lived up to their journalistic objectivity.

One said this was one of the most challenging activities she had ever done.

"Excruciating," said another.

Later, as we packaged their boxes for overnight delivery, Marcy and I were amazed at the variety of fruit. Some were perfect, with a slight "give" when we gently squeezed them and the color a rich, deep amber gold. Others were very, very hard, picked too green, and would never ripen. I was sure the editors would then blame me and my peaches for poor taste. A few of the novice pickers had selected extremely soft fruit, fine for today, but by the time they arrived in two days, they would be dripping out of the box, damaging the rest of the peaches by creating a breeding ground for rot. I wanted to fix their boxes and insert better ones.

"But you said they needed to take home perfectly authentic stories," Marcy reminded me.

PART FOUR

The Art
of Smell

Farm Smells I Remember

Crisp, clean air on a clear winter night. The early morning just before sunrise. Spring rains. Turned spring soil, breaking winter's crust. A neighbor's freshly cut field. Blossoms. Grape bloom. Burning leaves. Campfires in the orchard with my brother. Christmas trees when first brought inside. Laundry on the clothesline. Fresh sheets from that clothesline. Preserving peaches in a warm kitchen. Baked bread. Smell of wet barn wood. Wooden verandah that wraps almost all around the farmhouse. Dogs after running in the rain. The smell of dog on you after they shake themselves off in a splash.

Oil leaking from an old tractor. Stale air from deep well pump. Leather. Grease. Diesel. Cement and wet concrete. Sulfur. Pesticides. Herbicides. Saw dust. Rusting plows. Neighbor's cow pasture. Dust mixed with peach fuzz while cleaning up every summer evening after packing fruit all day. The scent of my dad — first when I was a child then later an adult. Everyday sweat from work. Smell of kids after playing outside all day.

Grape Bloom

I KNEW WHEN grapes bloom. Every berry on each bunch was born from a tiny, tiny flower in the month of May. We often scheduled some of our work according to blossom percentages, labeling jobs "done at 50 percent bloom" or "full bloom." But I had spent thirty-five years on a farm before I realized these tiny flowers also carry a scent.

The fragrance is subtle, easily unnoticed, and only detected close-up – I needed practice in order to pay attention. I first lift vine canes and stoop underneath the leaf canopy, settling underneath the cool shade, feeling the slight humidity as the leaves breathe, outside sounds muffled by the layers and layers of leaves. My eyes search for a bloom-ing grape bunch, the little flowers appearing like miniature stars clus-tered tightly. If some have already completed the flowering, my touch sends the old petals tumbling, like pixie dust showering the air.

I cradle one in the palm of my hand, moving closer and closer and noticing the variation in a single bunch, the bloom beginning on the outside shoulders and working its way inside. Tapped with my fingers, the berries shake and tremble; I gently exhale over the bunch and the delicate dust dances in the air. With my nose inches away, I take a quick breath. The fragrance almost eludes me; I can detect just a hint of the scent as I inhale slowly, conscious of my breathing, feeling the beat of my heart. Still the perfume is subtle. I push closer and breathe again and back off, as if I'm ready to give up. Only then does the aroma announce itself.

I can't quite describe the smell. I can't think of a proper metaphor. I can't create a label, nothing fills my thoughts. I have no memory of anything like this. I lean over and take a second and third pass. Words like "gentle" and "pure" come to mind but nothing describing flavor or fragrance.

I cradle another bunch and it has a slightly stronger scent, different yet the same. Subtle. I stop and close my eyes, slow down; the aroma becomes sharper, distinct. I crawl out from under the grapevine; the bright sun forces me to blink. My hands carry the fragile scent. Quiet, like a silence filling emptiness.

Grape bloom reminds me of farm smells. Usually not overpowering, often a blend of an earthy and musty aroma. Many things smell old on our farm. A barn with dust decades old, home to farm work animals, horses, mules, and many farm dogs. A milking parlor where a leaky tractor now sits and diesel and oil mixes with straw and leather. A farm with scents that match the seasons, the delicate peach blossoms and freshly turned earth of spring; irrigation water moistening parched fields as a shovel slices into the mud; dry autumn leaves left in the shadows and decaying when the first rains begin a natural composting; the dampness of fog as a year ends and begins with the work of humus in the good earth.

Each scent carries the baggage of memory and triggers a story and emotion. Unlike the world of promotion, advertising, and product identification which attaches a single scent (lemon-fresh!) to a product, farms generate a complex fragrance accompanied by equally dense memories. The freshly mowed cover crops remind me of my nearby small, rural elementary school. After the grasses were cut, we'd run on playgrounds with grapevines marking the fence line, chasing baseballs into weed patches of puncture vines and cockleburs, playing with classmates during the school year who worked our fields as farm workers in the summers. My farm's compost pile triggers a childhood remembrance. I watched our dogs and practiced sniffing the autumn air, trying to determine the direction of a scent and the origins of the manure and compost a fortunate farmer was spreading after harvest, having plowed his profits back into the land following a good year. (All through winter, that farmer will wear a broad grin.)

Grape scent reminds me to appreciate the simple things. The slowness of my recognition invites me to linger. Thoughts wander. I think of the scent of a woman long ago, a girlfriend's perfume that catapults me back in time to an emotionally charged moment. I was young and the memory was real, vivid, and honest. Still, wandering in a crowd, a whiff of the familiar triggers a story of an age filled with wonderful innocence.

Do California vineyards smell different from French or Italian? Aromas specific to a place? I wonder about the scent of Cabernet or Merlot and if there's a relationship with quality, year to year or even field to field. A blend of earth, landscape, and climate. Perhaps even the scent of those who work the land? My whimsical muse flowers while I'm working slow enough to smell. I pause to enjoy the moment, breathe the fragrance of grapes, smell the rosé.

Mud Time

I F I WERE part of the earth, I'd want to be mud, because of its aroma. Breathe in deeply. The rich smell of the land. Fat. Thick. Rich. A scent of life.

Mud time occurs three times a year on our farm. First, with the early-spring rains, the winter crust softens and transforms the hardened earth into a transparent aroma. A sense of optimism fills the air, and in the ensuing mud, life awakens with the first wildflowers and weeds. Spring mud is a harbinger of a new cycle; it marks the transition between the dormant season when mud is used as a convenient excuse to avoid work, and the beginning of the work season when the farm and its mud become demanding.

Then I look forward to the showers of autumn. The dust and sweat of summer harvests are replaced by the fresh scent of a storm. I can smell Thanksgiving coming. The tired, worn soil quickly soaks up the moisture. In the low areas where the rain puddles and pools, the end of the year announces itself in the stillness of a miniature pond. I see the reflection of clouds and a pale, clean, crisp blue sky with sunlight peaking through. Gone are the tracks of the past season, washed away; new footprints mark the earth and are left behind. I then carry the fresh mud on my boots everywhere.

Finally, in between these two seasons, I create mud myself whenever I irrigate my fields during the summer. It's a litmus test for the working of the earth: add water to the soil and stir until the fragrance meets your nostrils; take a handful of earth and squeeze it to measure water content. If it crumbles, it's time to water; if it remains a damp ball in my hand and holds shape, I'm granted more time and a few more days before vines and trees grow thirsty. This becomes my recipe for "in-season" mud time in order to produce luscious grapes and juicy peaches. Without irrigation, this arid land would revert back into a desert, with little mud time. Yet Thompson seedless grapes, which will be dried into raisins, and peaches, native to the highland deserts of China, thrive in this climate. When I provide a drink of irrigation water, my fields grow greener and the fruit fatter. When I smell mud, I can detect the scent of the approaching harvest.

But mud, like the land itself, is too easily ignored. After the initial rains or first irrigation, I'll forget to fill my lungs with its aroma. I become preoccupied with the chase of harvest. My senses stop exploring the gentle nature of mud, a nuance too subtle in a world moving too fast to take notice. Even at night, I sleep with my windows shut, blocking my opportunity to smell the earth with my dreams.

At times I find myself imposing my utilitarian values when thinking about mud and the farm. I catch myself calling my property "real" estate, a commodity with no life, its worth limited to dollars and cents. I plot a strategy focusing only on the short-term productivity that I can squeeze from the ground. Since I'm on this earth for only a short time, I treat the land as though it too has a transient character and a limited timeline of usefulness. Only the present matters; use up the land and move on. If necessary, deficiencies can be fixed with synthetic fertilizers. Mud is then redefined as a passing inconvenience; the real work commences once it dries out. I worry when I feel myself thinking this way.

Mud doesn't seem to belong in our sanitized commercial culture.

Even the rich fragrance of mud never seems to last long in my mind. Because I haven't trained myself well enough to maintain a sense of smell, I've found very few terms for scent, and without the right words, an aromatic moment is easily dismissed and slips by unrecorded.

I take mud for granted, as if it will always be there. I recall the drought years in California during the mid-eighties when rainfall totals were far below normal and snowfall in the neighboring Sierras was bleak. We had little snow that could melt and feed the spring runoff that would water already parched fields. Mud from a shower or storm then became a welcome sight, because it meant saving expensive pump water. For the last decade, though, we've had better rainfalls and deep snow packs, surface irrigation ground water is not a worry, and underground water tables have risen. Perhaps because I can now irrigate most anytime, mud becomes so commonplace I forget its value. I cease to notice the smell of wealth in my fields.

As I disk in early spring, something in the soil is plowed into me. I'll stop to study the turned earth, sharing the passion of gardeners who want to dirty their hands with the first hint of warm weather that breaks winter's crust. We plunge our hands into the moist ground and feel the damp humus. The touch of spring mud penetrates our flesh, and hours later, we still carry in our pores the renewed scent of a place.

In the arid heat of summer, the parched land greets me with dust. I create my own mud when the fine particles mix with my sweat, a blend of hard physical work and a struggle to keep up with the ripening peaches and grapes. As I wipe my forehead with the back of my hand, a streak is left behind on my face. The land mingles with my body scent, and I wear the resulting fragrance as a reminder of the fast-approaching harvest. Healthy market prices combined with luscious fruits lift my spirits, a bouquet of accomplishment, rewarding and full. Poor quality and the resulting low prices undermine my pride in my work with the land and sink my emotions; the smear carries a rancid odor that will not go away until the season is over. A lousy market drains my energy. I'm overwhelmed by the realization that no one wants my produce, my labor deemed worthless. The smell of failure fills the air, its stench penetrating my relationships and eating at my passion for this land. By the end of those long, hot days, no matter how many times I clean up, as I wash at the back-door sink, the rinse water will still cloud with dirt, the grit of the earth deeply embedded in my skin. I splash the water across my face, then pause and close my eyes, hiding with my hands covering my eyes, my palms up to my nose. Farmers must learn to live with the scent of their own mud.

I value mud time and its accompanying aromas because I'm forced to slow down and take notice of a seemingly insignificant fact – mud is transitory and my best-laid plans depend on things out of my control. An extra one-tenth of an inch of rain can be sufficient moisture to germinate an army of weeds in my mud. Cool temperatures keep the smell of mud around for an extra week and prevent a tractor from entering fields that need to be plowed. A heat wave will harden the mud into a baked shell, and I'll have to race to work up the ground before disk blades begin to bounce instead of cut. The changing nature of mud teaches a lesson in humility – I only have a small window of opportunity to work on this earth.

When I was growing up on a farm, lots of things got stuck in the mud. Tractors. Wagons. Disks. Shovels. Shoes and boots. Sometimes I grew frustrated about the lost time, or I would grow fearful that my father would be angry at me. But my father never got mad, and in fact, he always pulled me out of the mud without a rebuke. The last time I got stuck, I stopped for a moment and felt the damp, moist earth, breathing in part of the world that matters and remembering how as a child, I played in the mud – and grinned. In mud I see the true value of the land – it will outlive me and feels eternal, secure, stable, like home, like family.

Smell of
Workers

Iɴ ᴛʜᴇ Cᴇɴᴛʀᴀʟ Vᴀʟʟᴇʏ of California, peaches bloom in February. By the middle of April, the petals have long fallen and the teardrop-sized fruits have grown rapidly; they now look like small green oval eggs (pheasant, not chicken, eggs, about an inch long and half an inch wide) lined up along each of the skinny peach limbs we call "hangers." Hangers are about a foot long, and most will bear a half-dozen immature fruit. Once every few years, the fruit set is extremely heavy, and the miniature eggs line up like kernels of green corn, tightly packed together, already squeezing each other and fighting for space.

Space is what a farmer wants along these branches. Enough room for the fruit to grow fat, gaps between peaches so that each can swell and get big. We sacrifice a lot of little fruits for a few big ones. Small peaches never make it to market – people don't want one-inch-diameter fruit, so small you're buying mostly pit and not a whole lot of meat. So every year farmers knock thousands of fruit from each tree, thinning the crop down to a reasonable size.

"*Deje menos fruta*" – Leave less fruit, I tell the workers with my rudimentary Spanish. I wish Dave, my labor contractor, was around –

his fluent Spanish bridges the gap between farmer and worker. Since Dave's field crews are scattered throughout the valley, he also helps me see beyond my eighty acres, passing along information to assist my decision-making or offering his opinions to provide perspective, especially observations about crop size, what problems other farmers may be encountering, and predictions about the upcoming summer labor supply and demand. Dave is skilled at knowing how much farmers want to know and which farmers trust his ideas. I want Dave's help, and when he comes to check my workers, he adds, "*Queremos dejar menos fruta*" — We want less fruit, and the change of a few words of instruction could make the difference between small, undersized fruit and a good year for me.

I hold up four fingers — "*Mas o menos cuatro por favor,*" meaning I'd like at least a spacing of four fingers in width between each fruit hanging on a branch. Of course, every worker has a different hand size, so *cuatro* means a lot of different things, but it's a simple way to explain my intentions. Besides, in a perfect growing season a "small-hand *cuatro*" will be fine, yet the next year, with a cool or wet spring, a "big-hand *cuatro*" could have been right. So I guess while holding up my four fingers and shaking my head and shrugging my shoulders, utilizing my favorite phrase of approximation, *mas o menos*. I sigh as the syllables roll off my tongue.

Soon I can hear the dropping of fruit throughout the orchard. At first softly, as they descend into the soft, damp earth, or with a tap as they knock against the aluminum ladder each of the nineteen workers has. The field sounds like a light rain, then a sprinkle of "plop, plop, plop" as green fruits strike the metal or "thump, thump" as they free-fall to the earth. On the first day it's hard to get used to the sound. I keep thinking of profits falling from the trees. I look down and see the ground covered with small fruit, tens of thousands of fruit after only a few hours of work.

Tomaso, the field foreman, advises me not to look down. "Look up, look up," he encourages me, motioning me to keep my head up. "Up here, that's where the summer lies."

Away from us a few trees away, a quick chattering in Spanish and a worker is waving his arms, then he stoops over, gathers something, and proudly marches down the row, cradling four short logs. In my field are scattered odd pieces of wood. Some are skinny branches, remnants from the winter pruning; others are fatter, a young limb sawed off because of its location or odd angle. A few thick pieces of wood were also left behind by accident after I had grafted a few trees, renewing an old tree with new wood, hoping roots were strong enough to sustain young growth on top. In order to graft, I had to cut down the old, half-dead wood, which was hauled off in the middle of winter to the burn pile. I apparently had left behind parts of a limb, which were now being gathered.

The assorted pile of wood grows, and Tomaso pulls a few sheets of newspaper from his van and piles some kindling in the form of a teepee. He crushes and twists the newsprint into the form of a torch, flicks a match that bursts to life. The fresh scent of a flame slithers toward me as a single wave of smoke, a clean, crisp smell. Tomaso shoves the ignited paper into the heart of the kindling, and the smoke rises and dances in the still air. Then he adds more dry wood, and as the first limbs crackle, flames begin to dance. He fans the pile with his white straw hat, and a fire is born. He looks up to me, grins, and announces, *"Para lonche!"* — For our lunch!

Lonche, a midmorning break, a quick snack, part lunch, part rest. In the spring, the workers start here at 6:30 a.m. and thin about three hours before taking an extended break. I understand their rhythms, starting early, completing a few trees with a sense of accomplishment, then they take a proper break. On cool mornings the fire offers a break from the chill, but most important, it provides heat for their morning meal.

At about nine in the morning, from one of the pickups appears a simple metal grill, orphaned from a barbecue. The workers stop and trudge to the avenue toward their cars, where paper bags of food and drink sit. Quickly, tortillas are tossed over the flames, toasting on the steel rack. On one corner sits a can of corn, heating over the embers. On another corner appears a small frying pan with a mixture of beans, strips of shredded beef, and salsa. A few have burritos wrapped in aluminum that are heating over the flames. Some potatoes in foil are tossed directly into the fire, and sparks fly upward with the motion, bringing a chorus of comments and excitement as the men gather around the fire. There's talk of *pollo* but I don't see any chicken today.

I'm offered a burrito and tortilla but politely decline. I know I offend them but feel awkward about taking food, knowing these men work hard and don't make that much money. I head off to my truck and other farm work. They've done a good job with their thinning so far, so I want them to enjoy their break without a *patrón* hanging around. But I fool myself. I don't think it makes a lot of difference if I'm there or not. They know how hard they've worked, know they've earned a good break. As I climb into my truck, I can smell smoke in my clothes and see it hugging the orchard floor, lingering for a moment in the morning stillness.

An hour later, I pass by to check on their work and progress. By now a huge pile of wood sits near the embers, much more than they'd need today. I inquire, and Tomaso explains, "*Mañana*, you never know, maybe no good wood." But he is wrong; my fields are littered with lots of shredded branches and chunks of thick limbs for fuel. My land has been called quite "messy" by some, but I like the mixed assortment of textures. I understand Tomaso's situation, though. Other fields are very clean of debris, making a fire all but impossible.

Another worker lumbers over to us, arms full of a rich find of good *lonche* wood. He sheepishly drops it onto the pile and scampers back to work. Yet another arrives with three fat logs and unabashedly packs them into his pickup truck for another day; he then grins and announces he's helping me clear all this wood from my fields. Other workers howl and laugh. Tomaso pretends he didn't hear.

Near the fire, the senses come into play. The wood pops to a natural beat, the sap squeezed from the green pieces. Gradually I realize how warm one side of my body has become, and in tandem, Tomaso and I turn to warm the other.

But I enjoy the aroma of the fire most. It's relaxing and will stay with me for hours. I can close my eyes and be transported to memories of other fires. A limp fire my brother and I made when we camped out in the orchards – the leaves smoldered and generated more smoke than heat or light, and the smell didn't wash out of our clothes and blankets for weeks. The beach gathering with family when we occasionally joined other Japanese-American farmers and their annual spring clamming expeditions. The families circled around a beach bonfire, laughing and sharing stories, faces brightened by the flames and breathing in the scent. I felt we could be on other shores of the Pacific, three thousand miles away, as I squatted in the sand next to the embers, smoke getting into my eyes and nose, listening to the old Japanese immigrants chattering in their country dialects punctuated by long deep sighs that echoed across the ocean.

Tomaso leaves to monitor one laborer who has started a young, immature tree. "It's not strong," he observes, and advises, "Don't leave too much fruit."

As I begin making my rounds, checking the workers and their morning's work, I keep wondering why they made such a large fire. The sun has risen and all the men have shed layers of clothes, now hanging on various trees as if to mark time and progress. Dad once said that the best workers wore the least clothes – they made up for the chill by moving faster and "getting the blood to circulate." I imagine the slow workers were the first to huddle around the fire and the last to return to the cold fields.

I struggle to evaluate the work. I stroll and look upward, looking for branches with the right amount of fruit, checking to see if the four-finger spacing is uniform, hoping they've adjusted to the weaker trees while leaving more fruit on the larger, stronger ones. The slowest workers often don't do the best job, leaving clusters of peaches on a single branch, and months from now they will swell like miniature grape bunches, tiny fruit worthless for me. Are these the same workers sent to gather firewood? The pile didn't grow by itself; it took some time to hunt and collect. I began to calculate the price of these daily fires.

Then a scent teases me. A whiff of something familiar. The breeze carries an aroma like a faint siren enticing my memory. Then it's gone. I look around, perhaps for part of a discarded lunch as the source. I can smell grilled chicken. I turn and see wisps of smoke between trees and hovering by the workers' ladders. As they thin, many of them are chattering now; they seem excited by the new aroma, perhaps like the smell of good coffee in the morning. Grilled chicken takes me back to my childhood. I don't have fond memories of backyard summer barbecues with a father proudly standing over his art. Farmers work too long and hard. Summers meant harvests and long evenings when you could work outside until 8:00 p.m., and with a full moon, it was possible to see until 9:00 p.m. Dad often squeezed in another hour of work to take advantage of daylight. Harvest time was not barbecue time.

But my memories of the aroma of chicken sizzling over an open fire came from the Japanese community picnics in the spring. Families collected at a park or foothill location, pausing from the spring work, bringing bento — portable meals of sushi, nigiri (rice balls), and pickled vegetables. Many immigrants welcomed spring the same way, the entire community escaping the daily life and for a moment perhaps recreating a village festival from their homeland, flavors and aromas dancing in the breeze, lingering with the celebration.

Our main course of chicken would be made together, beginning the evening before when individual families took home portions to be marinated overnight. The next day one group arrived early to make the fire. By the time others arrived, the embers were red, the heat intense, and soon chicken skins would be searing, blackened lines etched into the flesh. I remember smelling the feast from far away as we children played games, then slowly gathered as the fragrance grew intense. We salivated and eyed our favorite pieces. I enjoyed the blackened parts, their deep smoky flavor. I believed the aroma could be licked off my fingers.

A worker starts to sing a Mexican song, and I once again hear the sound of thinning – "plop, plop, plop." I wander over to their large fire and grill and see four chicken breasts sizzling. I smile and breathe in the scent, fanning my hands toward my face, feeling an aromatic memory.

I turn to check a few more trees, and when I return, someone has flipped over the meat; now juices trickle from the crusty skin and drip below onto the embers, each hissing and sending up a barely visible wisp of steam and smoke, with an announcement to the workers drifting in the gentle breeze. I wonder when they'll eat these and who they belong to. Didn't they just have a break? Then I think – four breasts. For nineteen men? Surely there's not enough for everyone.

Tomaso appears with a long stick and gently nudges the blackened oblong objects from the heart of the fire. The potatoes must be done. I then realize this meal is communal, the crew will slice the chicken and share. The lives of farm workers are much more communal than most. They often buy food together, some live together, remaining invisible in a foreign land. But today when they eat together, the aroma of their meal makes them visible for this one morning, at least for me.

I'm not invited to participate in this meal, nor do I want to. I leave not wanting to know how long this extra break lasts. The scent of their fire has rekindled memories to last the rest of the morning. For the moment, it seems worth the extra break.

The Art
of Sex

"How long will this take?" asks Nikiko. Fifteen years old. A teenager. No longer a farm girl I can hide out on the farm.

I sigh out loud.

"No, I mean — I want to do this, Dad."

"Not too long. Ask Mom — she's done this before."

Marcy looks at her daughter and shrugs with that "just do what Dad says" type of response. They both grin. They're ganging up on me.

"Do I have to work?" Korio asks.

"Yes. The whole family."

We toss supplies into the truck, and the kids battle for prime wind-in-your-hair positions back on the pickup bed. As we drive the short half mile to the Sun Crest peaches, I can see them bouncing up and down in my rearview mirror, laughing and giggling with each jolt, and when I try to swerve around a pothole a moment of panic flashes in their eyes. Nice way to start work.

As they scamper off the back, Nikiko asks. "What's this all for?"

"OFM," I answer.

She twists her face.

"Oriental fruit moth. We're treating for OFM."

"Because we're Oriental?" Kori blurts.

"Oriental fruit moth. Little red worms that eat up our peaches."

OFM emerge in early spring and crawl into a shoot, gorging and eating the young leafy growth. Later as moths, they lay eggs, which hatch in May, just in time for a summer feast of peaches or nectarines.

"No one wants to find a worm in their fruit," I solemnly state.

"Better than half a worm," Marcy chimes and the kids laugh.

We each have a leather pouch tool belt, the type construction workers use for hammers, tape measures, and nails. The equipment makes me feel important, like strapping on a uniform and holster. I hand out the eight-foot plastic PVC pipe poles as if I'm a drill sergeant issuing weapons.

"Lock and load, private," I say to Kori.

He stands at attention and salutes. "Yes sir."

"Just open the buckets," Marcy responds.

Small white plastic strips, about an inch and half wide and three inches long, are layered in the bucket. Each has a black dot about three-quarters of an inch in diameter made from a slightly different type of thin plastic that I believe has tiny, tiny holes allowing it to breathe. A hardened plastic hook is stapled to the top of each strip. We reach into the bucket, pull out handfuls, and jam them into our tool belts.

"One for each tree. And keep the plastic hooks up, all in order."

"You never explained that," Niki says, trying to straighten out her bulging supply.

Mom comes to her daughter's defense. "In fact, you never explained any of this."

They're right. I keep thinking of how I learned most things on the farm — mainly through repetition and about the tenth time I finally figured out the meaning of my task. I suppose slow learning curves are out of place today.

"These are called pheromone strips," I explain. "Each strip has — how can I describe it — each has a type of perfume in it."

Niki glances up, eyebrows high. "Perfume?"

"Yes, perfume. A perfume that takes care of OFM."

"Does it smell good?" asks Kori. He brings one of the strips to his nose and looks up. I nod and he takes a sniff. The smell is sharp, not quite sour but with a bitter edge. He jerks his head back.

"Is it bad for us?" asks Niki. A quick look of panic crosses Kori's face.

"Only if you're an oriental fruit moth," I answer. Kori wipes his nose with his sleeve. "It's safe for us. It's not a poison chemical."

In fact, industry-wide, other chemicals haven't been working too well. Oriental fruit moths often take only a single bite of a peach that's been sprayed with some pesticide. They stop eating before it can poison them; they're known to throw up.

Kori looks like he might throw up. As we stuff hundreds of strips into our belts, he stops and starts to saunter off into the fields, where his PVC pipe will be magically transformed into a lance or staff. Then he'll start crafting castles out of the dirt.

We each take a row of peaches, and I demonstrate how to hang one strip in each tree. I insert one in a notch at one end of the thin PVC pipe, and when nudged against a branch, the plastic hook snaps into place and the strip is mounted for the season.

"Hang 'em high," I say. We want the scent to carry throughout the orchard.

"To perfume them to death?" Marcy comments. Mother and daughter snicker.

"The pheromone doesn't really kill anything." I pause to make sure I'm getting this right. "These strips have a specific pheromone — the exact scent of the female oriental fruit moth."

"Huh?" says Niki.

She's in the row next to me, a tree or two behind. I skip over to help Niki catch up and take the opportunity to explain.

"Well, a female has this aroma and the males are attracted to her . . . ." This isn't exactly the family conversation I was planning for the fields. "These strips are full of that aroma. And you see, you get . . . birth control."

"Didn't you skip a few steps?" Niki says.

I insert another strip, jab at another branch, and the clip snaps into place.

"Not exactly. The strips fill an orchard with this scent – this perfume of the female – and then the male moth can't locate the female."

Niki stops and blankly stares at me. I can't tell if my explanation made sense or she is teasing me. "Go ask your mom," I conclude.

A few late-blooming blossoms are still clinging to the trees. Green shoots peek out from branches. The spring sun is out, the air cool and brisk but no longer cold. I enjoy the opportunity for all of us to work together as family. Nothing special about the day – we're simply in the same place at the same time, working and talking together. Just us. And the oriental fruit moths.

"A variation of the rhythm method," says Marcy. "All about timing. In fact, nothing happens. No mating, no eggs, no worms."

I add, "You see, Niki, these strips prevent males from finding females." I load another strip into the notch. "The orchard is flooded with this scent, this perfume is everywhere." My free hand waves across the field. "And the males just can't find a partner."

"Oh, confused males," Niki concludes. "That's nothing new."

Marcy bursts out laughing.

Never quite thought of it that way. Did recall that in promotional materials, this treatment program is called "confusion method."

"Yea, I guess you're right. Confused males."

"The scent is everywhere. It fills up the fields," Marcy repeats and mimics my hand wave, panning across the horizon. Both females are giggling out loud now. "Actually, Niki . . . It's frustrated males."

Both laugh, almost uncontrollably.

"Like ultimate safe sex!" says Niki.

"Yes! Exactly! The safest," Marcy exclaims.

For a moment I'm pleased. Learning is taking place right here, out in the fields. Finally at least someone is taking this seriously.

"Wait," I realize. "Niki, you know about safe sex?"

We continue to walk and work in silence. We're each in a rhythm clipping a pheromone to each tree. I work a little faster than the others, pulling ahead by a few trees. Kori is content in another part of the orchard; he's begun collecting rocks in his pouch. Meanwhile Niki hums and Marcy continues to giggle. They work next to each other and every once and a while I can overhear a phrase like "sex in the fields" followed by laughter.

At the end of the row, I turn to help them complete their rows. We all start new ones together. Niki asks, "Is there a human pheromone?"

"Don't know. Go ask Ms. Expert over there."

Marcy grins. "I suppose so. I like to think it's part of charisma." She laughs again.

I say, "If I knew for sure I'd make a zillion dollars. Imagine controlling that market, selling the creation as a perfume, an aphrodisiac, a lure."

"Aphrodisiac?" my farm girl asks.

I stall, pretending to have trouble.

"Something that stimulates desire," Marcy then says.

Niki ponders the thought and quips, "Don't pheromones act as a type of repellent? It repels because males seek too hard?"

I brighten with the new concept, a contrarian theory of marketing pheromones – "It distracts because it attracts" – from the point of view of an overprotective father of a teenage daughter. "The more he tries, the more he dies." I'd market this as a high-tech rhythm method and my target consumer could be fathers – precisely at the moment when we discover our little girls have grown up to be pretty young teenagers.

I shudder with the reality. I smell a fortune.

Farm Aromas

IN SEPTEMBER, my nose takes me into the raisin vineyards. I follow the scent of green grapes slowly drying into the blackened morsels. What's the smell of grapes exposed to the elements for those twenty days? Initially little. My eyes study the thousands of paper trays lined along the vine rows, in the direct sun, a bright green carpet of grapes. A few bunches are yellowed and riper, most a neon green for the first few hours, reflecting the brilliant summer sun before they start to dehydrate. In a day they will all yellow, quickly losing the sharp hues and fading into a dull amber; they haven't decided whether to stay a grape or transform into a raisin. Gradually a watercolor lavender hue appears, starting with the very top berries most exposed to the heat, a sun worshiper tanning on one side first.

That's when the aroma begins. A sweet smell of curing. Transient – the slightest breeze will easily dissipate the odor this early in the drying process. Only in the evening as the cooler air settles along the surface can I breathe in deeply the scent of raisins. With each passing day it intensifies. The top layer of berries darken first, toasted as if in a broiler; then the bottom layers gradually color into a light violet. Wrinkles appear on the skin as the berries shrivel. I'm not sure exactly how moisture is lost in each berry. One scientist described the surface as like fish scales that slowly open and allow water to escape, leaving behind the fleshy meat and sugars. Another refuted this and claimed the grape skin is porous, like that of all fruits and vegetables, and with heat and a dry environment, moisture is transferred from internal to the external. The reverse is true too – that's why with a rain, raisins absorb moisture and become bloated. I imagined the surface to be filled with tiny, tiny holes, water escaping through these outlets. And aroma too.

The caramel, sweet smell of a raisin seems to come from the inside; it's not the odor of the skin drying. The rich flavor of raisins is also in its scent, each berry like a pressure cooker with the contents patiently transformed by the late-summer heat of one hundred degrees. The meat of a grape follows an ancient recipe of "stewing" and "simmering" for twenty days inside the skin, gradually releasing moisture and fragrance.

I am drawn into a vineyard and silently enveloped by the perfume. This annual migration into my fields is a journey into a place where I'm led by a scent. The rows of vines and their trellis wires create a green wall on each side of me, capturing both the aroma and myself. This archaic method of creating raisins by simply laying grapes on trays in the sun speaks with a language of the past. Slow. Curing. Exposed. Both the raisin and the raisin farmer.

The spring El Niño rains are felt in September. The cool, wet months of February and March with double the usual rainfall, coupled with the intense showers in June, when we normally just get a trace, combine to make for a late-maturing grape crop. Sugars don't begin to accumulate in the berries until late August, and we won't be picking for raisins until mid-September, a full three weeks later than normal. With the approach of the equinox in late September, the days grow shorter, temperatures slightly cooler, and farmers become fearful. It will now require twenty-five to thirty days to dry this crop (instead of the typical twenty), pushing us closer to autumn showers. The lament of the old farmers echoes throughout the homes surrounded by vineyards: "The weather in September is not to be trusted."

Three brothers meet to talk about the El Niño grapes. All three have fine jobs in the city, where two work for agriculture-related businesses. But they have inherited the eighty-acre farm from their father, who died a few years earlier, and the siblings still gather for an annual family meeting prior to harvest. In the final years of the patriarch's life, the sons met to make sure the harvest would be done efficiently and on schedule. A sloppily picked raisin tray angered and frustrated their aging father, and for weeks he'd trudge through the fields, grumbling about the lousy job while bending over to even a heap of grapes carelessly piled atop each other. So the sons helped coordinate the pickers, the turners who flipped each tray so that the underside of grapes could dry, and the boxers who helped bring in the crop and transfer the thousands of individual trays into boxes or bins. The first years "without Papa" were melancholy but normal in terms of weather. This year is not.

The spring rains have led to damp fields for weeks, perfect for the growth of mildew on the grapes. By the end of summer, berries are splitting and rotting. Driving through the vineyards, a farmer can detect pockets of stench in the air as bunches ooze with juice, fruit flies hovering over meals and molds growing and sporulating. Making raisins will be difficult this season. With the predicted lateness, the damage can spread; a 10 percent rot problem could easily explode, jeopardizing the entire field.

The brothers have an option. A winery they have contacted and negotiated with will buy the grapes, and if they are picked and crushed soon, the rot will not damage the whole crop. By selling "green" they can possibly save a lot of headaches. But for the grandsons of an immigrant from Armenia, whose father faithfully made raisins every year of his adult life, this marks the first time the Shapazian family will not have raisins in the autumn. Mark, the eldest son, describes his father as a stubborn man who was strong until the day he died. His father told stories about "staying the course" and how making raisins was something the family could count on through the good years and the bad. "Some things don't change," Vaughn mimics his father's voice. "We make raisins. Always."

The sons are torn, a modern-day struggle of balancing tradition and family with the realities of making a living. Perhaps in their father's era, they made raisins because they didn't have many options. Years ago decisions seemed easier. The brothers keep asking themselves the same question that refuses to go away: "What would Dad say?"

They debate about the quality of the raisins, bunches that are lost to mildew, berries damaged with rot versus the quality crop the family always seemed to produce – the fat, sweet raisins of the Shapazian family are well known. The sons take pride in maintaining that legacy. This year, though, the berries look weak. The abnormal wet and coolness delayed maturity; even the skins look dull. "Going green" would be easier and simpler and safer.

Another son, Robert, recalls, "We had a meeting in the old barn. We smoked cigars and talked about breaking tradition. We agreed not to make raisins."

He then relates a sense of both relief and yet an odd feeling of his father's ghost sharing the decision with them. After a few minutes with their cigars, they concluded, "You know what Dad would miss the most this year? He'd miss the smell of raisins in the fall."

I think their father would also miss the smell of risk. Absolute risk. I walk down a vineyard row during the final days of summer. If the weather has been kind, the heat lingers well into the evening, the season refusing to give in to the earth's orbit, still claiming to be summer. The ground remains hot, radiating heat – I swear the raisins cure from the ground up. By the time I reach the middle of the row, I've stooped over dozens of trays, clipping out rot from the bunches, reaching over to snip a stray vine cane that grows away from the trellis and casts a shadow over the half grapes/half raisins. I grow too accustomed to the sweet scent and have to pause in order to pay attention and recognize the aroma. As the sun sets, the inversion layer of air forces down the perfume, and I'm surrounded. I wonder if it can permeate my clothes, like tobacco smoke. If so, I can't detect it.

The millions of individual berries can fill my air space only because they lie open, exposed. There's a price I pay for this moment, though; I live for weeks paralyzed with fear of rain. Though I try to suppress it, I automatically turn up the volume during weather reports on the television or radio, "shhhing" my family and demanding quiet. A cloud of smoke on the horizon from a farmer burning tree stumps catches my eyes, and I stop to study the dark formation, verifying it is not a storm cloud. I don't sleep well, tossing and grinding my teeth so loud I wake Marcy. I leap out of bed with an odd noise at night as if I'm on call and a pager has gone off.

Raw nerves for nothing. Silly to think there's anything I can do. The grapes are green and literally too heavy with moisture. Trying to roll a wet tray to protect it from rain means not only do you have to unroll and open the paper again, but the grapes that need sunshine and heat will then be lumped together on the bottom, a heap with little hope of drying.

I believe only a few farmers manage to overcome anxiety and not worry about the weather. I can't, nor do I want to. The risk is genuine and absolute. A storm will damage the harvest, and at the same time I can attach no blame. There will be no finger-pointing or accusations. The two sides are clearly drawn: the farmer versus nature. The smell of this type of absolute risk is clean and somehow feels pure. Naked and bare.

Sometimes I feel like the last real farmer, the
boy who bets the ranch and plays a final hand of th
in the crop, drive the herd to market, and the score
exchanged for income. Lose and there is no harvest,
gamble with no oddsmakers fabricating chance, no human intervention affecting the outcome. No guarantee of return. Few raisin growers walk with a swagger. We live with a complete faith in nature and
a lack of confidence in the outcome.

The scent of risk is the same for the raisin farmer in 2002 as it
was in 1902. A shared memory as I pool my chances with generations
before me, the past matched with the present, a connection with
something larger during a moment in a vineyard. Dad and I walk a
vineyard, moving grapes out of shaded areas, leveling those bunches
that had been piled atop of each other, clipping out rot and snipping
the fat bunches in two. We work in adjoining rows; I can hear the
rustling of paper and a few grunts as he leans over to reposition a tray.
At the end of the row we pause and stretch aching backs, standing tall
again and rolling shoulders.

We're breathing hard, ready to call it a day, knowing we'll each
take another row as we work back to the barnyard. I watch him close
his old eyes; his nostrils flare and his chest rises. He's taking in the
smell of the raisins. I do the same. A son and his father. We rest. We
breathe. The scent of the father and his father. In raisins I smell family.

Dad keeps a secret collection of raisins in old coffee cans in the barn.
I think they are for afternoon snacks when he'll take a break from
welding or fixing something. Instead of going inside the farmhouse,
he'll pull off his gloves and grab a handful of raisins. I believe they
remind him of a harvest brought in safely.

While looking for nails or a particular-sized bolt, I pilfer through
dozens of old coffee cans, even a few with tin covers and sharp metal

dges from when we used a key to open the can and rolled a thin steel band off the side and near the top lip, breaking the seal of the coffee yet leaving a tightly fitted cover. In one of them I once found some of Dad's snack raisins. First I shook it, and instead of the clang of washers or bolts bouncing against the sides I heard a soft thud. When I twisted the top, a faint hiss told me this can had been sealed for a while. I lifted the lid and an aroma of ancient raisins lifted into the air. I had invaded a memory of Dad from long ago, a secret stash from a harvest many, many years before. Not quite opening an Egyptian tomb, but the emotions of discovering a treasure were similar — I was breathing the air and fragrance of Dad's youth.

I recall a conversation with Baachan when I returned from Japan in 1976. I had lived for months in her family's small village outside of Kumamoto. I slept, ate, and rested in the ancient farmhouse she was born in and grew up in before immigrating to America as a young woman. She had never returned since departing in 1918, and now Baachan seemed a little scared about a visit to her childhood home. She was certain much had changed.

My photographs shocked her. She couldn't believe the images of the old thatch roof, now covered with thin sheet metal but the thick and tightly bundled straw was visible along the eaves. The house still had no flush toilets; the old outhouse stood leaning against the main frame, as if resting on its final legs. Inside, the dark wooden beams stretched overhead and the kitchen area still had no floor, the earth compacted and worn smooth by generations of feet and a century or two of sweeping.

Baachan didn't believe her eyes. She stopped to wipe her glasses and studied the photos again and again. Then I told her about the smell of the smoke from the fire used to heat the water under the old *ofuro*, the metal bathtub that looked like an inverted bell. The steam rose from the liquid as I soaked and relaxed. Jiichan, my grandmother's brother, who took over the family rice farm in 1920, stoked the

flames and fed the embers, smoke gently rising around me and dancing with the steam.

I joked with Baachan about the occasional stench from the outhouse and the price we paid for having it so close to the house. Instead
of moving an outhouse when the pit beneath it became full, we had to
remove the contents with buckets and large ladles while trying to
keep the odor at a distance with bandanna over our noses and mouths.
Baachan smiled and even laughed; the stink must have touched a
memory.

Then I tried to describe the aroma of the wood. The timbers were
dark and thick, not in appearance but in their smell. The scent was not
like the freshness of a mountain pine forest, yet it was not simply old
either. The only way I could describe it was as like the smell of raisins
– cured and rich, as if over time the farmhouse had properly aged.

Baachan stared and blinked. She seemed to start drifting within
her thoughts, and her seventy-year-old eyes watered. We sat silently,
breathing deeply. I wondered what she was smelling. A few days later,
she told us she'd like to go back to Japan and visit her family and the
farmhouse she hadn't smelled since her childhood over sixty years ago.

Aromatherapy

Each September I perform an annual ritual – I drive the country roads with my car windows open and identify a passing vineyard by the aroma. It announces where I am – a fifty-mile radius around my farm just outside of Fresno, California, where 99 percent of all raisins in the United States are grown and made. I associate the smell of raisins with geography.

I used to have a sense of the local territory by the smells. Drive to the west and the small dairies of Riverdale and Laton carried the fragrance of cows and manure. Right after rains it smelled rank, but most of the time it carried an odor I became accustomed to. I'm sure the locals paid no attention. Their ranch houses sat next to milking parlors, pens, and barns. Only those of us who were sheltered and used to air without natural odors sighed out loud when we got our first sniff.

To the east, through much of spring the orange blossoms perfumed the air so heavily my eyes itched, and I could almost feel the air going down thick into my lungs. Bees from miles away were lured here as migrant workers performing the job of pollinating. I now associate the smell of oranges with an orchestra of buzzing sounds.

South, there were the open lands planted with alfalfa. Most any time of the year, I could smell at least one field that had been recently cut, or another curing in the sun, and yet another just baled. I now associate the term "fresh" with cut grass – breathe in deeply the color of green.

To the southwest near the Coast Range mountains, there was the strong smell of sulfur in the air. Once as a teenager, we traveled to a remote high school for a basketball game, and after our contest we showered and the water felt slimy and tasted bitter. They explained the water was rich with lots of minerals. We thought it smelled like rotten eggs, especially after we got whipped in the game.

Scent identified a place, part of an aromatic geography known by the locals. I used to know these smells, but they have become blurred in recent years. With an increase in population and the proliferation of traffic and cars, our valley has become a basin for smog and exhaust. The delicate natural scents, even some of the strong smells of manure, seem to be overwhelmed by these manmade odors.

And farming has changed. Larger and larger dairies must comply with air quality standards. Waste disposal and control of odor have become part of good farming practices. No smell is now a good smell for farm operations. More and more of the open lands are planted with higher-valued crops of tree fruit and grapes. The alfalfa fields have migrated to other regions of the state and have increased in size. The strategy to diversify a small farm with different crops – vines mixed with tree fruit interspersed with odd lots of open ground planted with row crops – these practices have been abandoned. That wonderful mix of drying hay next to the freshly disked orchard with a row of oranges along a driveway is now gone. Our smells are much more uniform.

I suppose what I can now detect is the smell of efficiency. Working smaller blocks can be taxing, difficult for large equipment and requiring extra hours of management. I sometimes curse at my short rows and the wasted minutes at the end of every row as my tractor swings around and equipment is lifted and then reset. During those moments I breathe in the fumes of wasted diesel fuel. Long rows have less dead time.

Yet as the economic challenges of small family farming multiply, the diversity in aromas seems to map a new strategy for survival. I wonder if the old model of mixed crops on mixed land doesn't help small operators get through bad markets for one crop, hoping it'll be balanced by stronger ones for another. I've learned that farming organically is benefited by a diversity in habitat for biological control, which creates multiple homes for good bugs so that they settle in year round and take care of bad bugs. Newer research investigates the role of multiple-cropping systems to create a broad spectrum of molds and spores to help combat the devastation of a single pathogen invading a monocropped field.

Friends who sell at farmers' markets are now divesting themselves of produce beyond just peaches, plums, and nectarines and adding new crops and new products to expand their selling season. Their farms now have a diversity of fragrances. Our farms can use a healthy variety of scents. We need to drive through the countryside and breathe in the value of difference. I call this aromatherapy for family farms.

Public Art

A GROUP OF bicycle riders from the Davis Bike Club contact us. They have read some of my stories and describe to me a wonderful program called "read and rides," where the bikers read the literature of a place, then spend a weekend cycling through the region, combining the story and images with the reality. They've read and ridden Steinbeck's Salinas Valley, Jack London's Sonoma Valley, and Mary Austin's "Land of Little Rain," the eastern Sierra. The idea intrigues us, and Marcy and I make a deal to host them.

Their tour normally includes a map of the area plus a narrative tour with places to pause and reflect on literary passages. For example, on the riding itinerary, at "Point #7 – King's River," about ten miles from our farm, they ride over a series of bridges that span the sole river that brings water to our fields. The organizers, Donna and her husband, Richard, remind the riders "to imagine Yokut villages in the area" and reprint a short Yokut Native American prayer. This is followed by my passage about the irrigation system that "brings life to the orchards and vineyards. Only with water will peaches ripen and the scent of bloom lingers in the air. The vine buds will push and the pale green of fresh growth emerge pure and delicate year after year."

We invite them to the farm for a light brunch and farm walk as I read and talk. During the morning I instruct the group to step silently and breathe deeply. The peach blossoms are in full bloom, and the air is filled with their delicate fragrance. The slowness of walking awakens their other senses, and the sounds of the farm and countryside acquire new life. But this group need no lessons. They are used to going slow, biking down country roads with a leisurely pace so they can feel the landscape. No instructions to drive with windows open. They know how to fill their lungs with the scent of a place. I hope my words and stories add meaning to what their noses detect. Damp earth. Wet wood. Peach blossoms. Wildflowers. Clovers. A breeze and the sensation begins again.

The bikers walk, stroll, saunter through the orchards with their heads up. A few lift their noses like an animal surveying the landscape. I wonder how my air compares with Steinbeck's or Jack London's.

The smell of a region, the aroma of an orchard, the scent of a field – they become public art, the expression of a place. Truly public, no ownership or control of these fragrances as they wander throughout the countryside, sometimes heavy and still, other times fleeing with the slightest breeze.

The story of this part of California and the meaning of my farm is conveyed via invisible narratives detected through someone's nose. Most of the smells are pleasant, light, and airy. Others are hard – like the crisp, clear frost that rips into a valley in early spring, blackening delicate green buds and leaving behind the faint smell of rotting tissue; or the odor of sweat mixed with dirt, grime, and ten days straight of hundred-degree heat – an earthy aroma of harvest work and the bringing in of a crop that drains and dehydrates the farmer. The meaning of my farm necessarily includes the challenges facing a farmer; risk must be part of our aroma art.

Scent combines with a rush of memories and generates a reaction to a place. Initial reactions to odors tend to be personal. The group begin to share stories about the places they've ridden and the rich smells they know. Because they ride in the open. Exposed. Available.

I share with them one of my biggest fears – that the average person often has no memory of smell. Aromas fail to touch because they have no meaning. I explain that when I open a box of raisins, I quickly put my nose to the top, and breathe in a farm and the smell of my family. Raisins are then to be eaten communally, a group sharing in the flavor and aroma while memories of Mom's kitchen and baking cookies dance in our minds. But for many, raisins smell no different from candy, the want for sweetness overwhelming all desire. Sugary becomes the aroma consumers seek.

I know of gardeners who randomly toss in a handful of herb seeds as they plant. As the year unfolds, while weeding, they're pleasantly surprised as a passing breath of dill or thyme crosses their path. Unexpected and sudden, like the best surprises from nature. The fragrance brings a smile no matter how tedious the work. Medicinal art.

I love watching animals stop to sniff the air. I assume they're smelling for danger signs and the scent of predators, but they also know places by their smells. The sweet fragrance of spring grass to a rabbit may mean dinner; later they all scurry for shelter with the smell of rain. Our farm dogs make daily rounds, going over for "coffee" at my folks' house a half mile away, dropping in at the neighbors for a snack or visit, sulking home late at night like an overdue teenager. Along their routes the dogs will frequently stop to smell the air, searching for the new scent of a stranger and detecting places they've spotted as they marked their territory. I sense they take comfort in detecting their own fragrance. I don't know if they get emotional about these aromas, but I have witnessed numerous times when the farm dogs pick up the scent of prey and grow excited at the possibilities of a hunt. They start yelping and frantically pacing back and forth, some deep emotional instinct touched. The smell moves them.

There's a moment when farmers pause at the end of a harvest season. For me it's with the raisins, when the last tray is picked up and the crop safely stored. I stand with my father next to the rows of square plywood bins stacked four high and towering twelve feet in the air. Our complete crop is in those one hundred bins of raisins, each crate weighing about a thousand pounds. The air thickens with the sweet smell of raisins. If I could bottle that aroma combined with the emotions of the moment, I'd be a rich man. The scent contains the smell of risk and harvest. And now with it safely under shelter, we can't help but grin.

Scent of
My Father

I GREW UP with Dad's sweat. Some days he'd come in for lunch or dinner soaking wet, dressed in a blue Penney's work shirt with a dark streak painted on his back along his spine and a damp V shape over his chest and down a few buttons. I could measure his day's effort and exhaustion by the number of shirt buttons the stain stretched down to — a one- or two-button day was long, a three- or four-button day must have been exhausting. His collar and armpits were not as moist, possibly dried out on his way home; yet telltale white salty streaks marked the ebb and flow of his laborious workday.

Dad's sweat did not reek. Instead, like that of many Asians, whose body odor seems less pungent, Dad's scent was more delicate. I remember Armenian neighbors whose fathers carried a strong aroma, like their presence — commanding and dominating. Sweat that shouted. Dad had a gentle smell, more like a whisper.

I did not learn of Dad's scent from physical contact of hugs and playful wrestling or even long handshakes. We rarely touched. But I did work side by side with him for hours, first as a child growing up on a farm, then as an adult who came back to farm. His sweat carried

a slightly salty tanginess, with a twist of lime or lemon. I always associate his smell with work, the aroma of hard, honest physical work.
Dad and I rarely tossed a baseball; I have few memories of actually
ever playing with him. Sweat was not recreation or sport. Sweat was
meant for work.

I could sometimes tell the work he had done that day according
to his scent. Curious and harsh smells came from times Dad had
applied pesticides. Oils used for dormant orchard sprays soaked into
his skin, and then he reminded me of a greasy Chinese restaurant
kitchen. Omite, a miticide no longer used and highly toxic, had a very
distinct, almost sweet smell. It even carried for miles in the country
air; I could tell when neighbors also sprayed. Sulfur dust had a burning smell, and no matter how long Dad showered, the scent must have
penetrated his pores and stayed with him for hours into the evening.

But natural aromas are what I remember the most. The strong
smells of chemicals, while overpowering at the moment, left little to
imagination; I could identify the origin of those odors with little
thought. Farm smells, though, stimulated the senses.

The smell of cut grass – I imagine Dad had spent the day shoveling weeds, probably johnsongrass with thick stalks and dense foliage.
At certain times of the year, if we sat close enough, I could detect the
nutty fragrance of the seed heads that had lodged themselves in his
pockets or folds in his work clothes.

The thick deep aroma of mud – in spring it reminds me of life
germinating and renewing itself; in the damp, cold days of winter I
think of short days with darkness before 5:00 p.m. and a wet chill that
coats leather boots and soaks into socks. I remember watching Dad sit
and pull off his shoes, peel off his socks, revealing feet that smelled
like mud, red with cold, joints stiff, socks stained with the aroma of
mud. After drying overnight, his boots still smelled like mud in the
morning, a crisper, colder scent after the night's frost.

Some days a smell of grease and oil clung to Dad even after washing up with soap and water. His hands looked dirty, with stains embedded in his fingerprints and a blackened layer lodged under his nails. On days when he welded, a startling metallic smell stuck to his skin, along with a tiny fragrance of singed fabric from specks of molten metal landing on his clothing. Leaning close to Dad, I could also detect a distinct odor of burned hair from wild sparks splashing in the air.

The art of smell, the scent of a father — who he is, not what he wasn't. Aromas marked where Dad had been and what he had accomplished. He was not the first to carry home a scent of work, nor would he be the last. Now that I do some of the same tasks, I hope I smell the same to my children.

Burning
Woodpiles

Ask: "What does a father smell like?"
My answer: "Like the smoke from a campfire."

IRONICALLY, I have never gone camping with Dad. We have never
spent overnight bonding trips together out in the wilderness. He
never taught me the proper way to build a small campfire with kindling stacked into a teepee shape or how to separate green and cured
wood. I'm not sure many dads really did these things anyway – in talking with my peers, most of us certainly don't know how to make a
proper cooking fire, and I suppose the world will have to brace for a
generation lacking in essential outdoor survival skills. Yet whenever I
smell wood burning, or stand for any length of time near a fire and
allow the smoke to linger and penetrate my clothes, I think of Dad.

In the old orchard that Dad and I planted three decades ago,
every year branches and limbs die. I saw and drag them to the ditch
bank, where for months they look like Monet's haystacks in the
early-morning light or late-afternoon sun. We burn the woodpile in

the autumn, when there's finally some time to finish a chore, a pre-cursor to pruning for the next year and a time of renewal.

Most small farms have to have a woodpile, a collection of limbs and stumps mixed with odd clippings and prunings from the fields and yard. A place to stack the dead — trees and vines that have aged, or the old branches that were sawed and hauled off when grafting in order to make way for the new. Annually our pile stands ten feet wide and twenty or thirty feet long, a stack eight feet high and weighing tons. Tucked away "out back," these piles tell of changes on our farm. It's a modern-day slash-and-burn method as the farm evolves. Burning is like a purification ritual as the old turns into ashes. Sometimes I'll spread the dust in the fields when the crema-tion pile rises high.

Dad and I head out early with pruning shear, shovel, and hand saw, along with some sheets of newspaper and a book of matches. Dad has a routine in order to start the fire. He will wait for the occasional day when the wind blows from the west to the east, in order to keep the smoke out of our eyes and sustain a natural fanning of the flames. First he'll find a very dry branch and begin cutting it into smaller pieces for kindling. He locates a sheltered cubbyhole along the west-ern edge of the pile; there he crumples the newspaper and jams it under the sticks, flinging a match and igniting flames.

Quickly the fire leaps into the pile, and the dry, withering leaves explode in a crackle. Initially the slender and well-cured branches catch fire, and the entire pile jumps to life. The smoke is thick and dense, the smell choking and heat intense. Wild moments burst forth. The initial rush of fire, racing to consume leaves and twigs, creates a few moments of anxiety. This fire is not to be controlled. It will grow and spread much too rapidly, and we will not be able to contain her energy.

But after ten or fifteen minutes, the sticks ignite, then die out; burning the thicker two-to-four-inch branches will demand harder work. We begin with the small pile of red embers. Dad piles on logs, smoke curls with a lazy rhythm, and the scent of the blaze settles into our clothes, hair, and skin. As the day unfolds, the flames carve out a cavity, a pocket of empty space where branches once lay. I push the pile inward, sometimes with the help of a tractor, closing the gap and feeding the fire. Ash dust mixed with sparks billows into the air, the new fuel crackles, and the inferno leaps to new life. The smoke penetrates my lungs. I can no longer smell the fire; I feel enveloped.

I'll spend the whole day with Dad, watching his methods, adding an unsolicited opinion that he greets with a nod. We talk little, listening instead to the hiss and crack of the wood and the popping and spitting sap. A stray piece of lumber is tossed in the pile; we stand back with a shift in wind and cover our eyes, blinking to get the smoke out. We flirt with adventure by jamming in too much wood at one time. The fire burns to new heights. The heat grows so intense we're pushed away, and we have to shield our faces with our farm hats, protecting skin from the spike in temperatures. We stand in silence and watch, a shiver of terror with the power of the flames, a humbling feeling before the fury. My imagination runs wild, and were Dad a God-fearing, zealous evangelist, I'd swear he brought me out to the woodpile to teach me a lesson. The flames soar and the timbers crackle and shake, then snap and tumble to the earth, sending sparks flying into the air as if struck by lightning, the Almighty as the omnipotent power. I'm really glad we're Buddhists.

Occasionally, the smell of burned rubber piques my interest. The aroma is strong, distinct, and nasty. I've learned to look immediately at my work boots — have I stepped on an ember? Or while feeding the flames I may have walked across the ash pile, still hot and hungry for fuel. Once I had to retreat and take off my shoes. I tapped the soles and they were so hot I could not keep my fingers on their surface. I felt my socks — they too were flushed. Had I kept my boots on, I would have burned my feet, my shoes on fire without flames, the scent of melting rubber my only warning.

Neighbors will lift their noses when they smell our woodpile, scanning the horizon as they look for the line of smoke filling the sky and wonder what's burning. Black, dark smoke, telltale sign of an oil-based fire. But the gray, almost white column identifies a farmer. A cluster of columns usually means an entire orchard or vineyard is burning; a farmer has pulled and piled something old while making way for something new.

What began as a huge pile early in a day ends with ashes. I don't believe Dad feels powerful with each burn; we don't work as conquerors laying our victims to waste. We renew the farm with each fire while reducing the pile to nothing. At the end of the day we trudge home, worn out, but the pile is reduced to ashes, the embers burning all night. Tomorrow we can roll the thick stumps into the center and they'll ignite and burn all day. Only later, at home, do I realize how smoky we smell. The scent penetrates our clothes, even our hair. We shower and wash our clothes together, and still the smoke remains, embedded in the fabric, and for weeks we relive the autumn burn pile, accompanied by a faint scent of smoke.

Dad and I are about the same size. Years ago when I lived with my parents, we occasionally got our clothes mixed up. As I buttoned a work shirt, I thought it smelled old with a hint of his sweat but still clean. Now, at nearly eighty years old, Dad has lost weight and slouches as he shuffles along; he no longer needs all his work clothes. He gives them to me, and even with much laundering, I can't wash out his scent — nor do I want to.

Dad and I talk about old schools that once dotted the countryside. Some abandoned buildings still stand. Rosedale. Prairie School. DeWolf School on the corner of DeWolf and Central. The Masumotos moved around; Jiichan couldn't own land because of the racist Alien Land Laws that specifically excluded "Orientals." New generations of Italian, German, and Armenian immigrants with heavy accents and alien cultures also arrived in Fresno, but they were allowed to buy farms, settle down, and plant roots. I imagine they learned the smell of one place. The Masumotos were farm workers, or at best leased land for a few years before moving on. Drifting across the countryside like a wandering scent caught in the breeze.

"Japanese were easy renters," Dad said. "They only asked for fifty-fifty split 'cuz they couldn't own land. Others could bargain for sixty-forty. Sixty percent for the renter, forty to the owner. Ten percent, big difference." He grunted low and long.

The last place Jiichan rented was out near Selma on Manning Avenue. They stayed there for a few extra years because the kids wanted to stay in one school. The grapes grew strong under the family's care. A widow owned the land and had no intention of farming it herself. "Japanese good renters, took care of their places," Dad remembers.

The plan was to keep renting for a few more years until the oldest son, Uncle George, had graduated high school and could work in the fields full-time. Everyone saw signs that the Depression was ending: raisins sold for forty-two dollars a ton in 1939, forty-nine dollars in 1940, and then jumped to fifty-six dollars in the summer of 1941 with a light crop, and there was talk of the price climbing higher. Uncle George was so confident that he went out and bought a new car. "Something reliable so we could drive to other farms and make extra money," Dad rationalized. I'm sure both teenage brothers enjoyed the smell of a new car.

The family could soon swing a deal to buy their own farm too. Dad was just finishing school and could add his strong back and quick hands. Then Uncle George got drafted into the United States Army. He left for training in the fall of 1941 and plans were put on hold. The family would have to lease land for a few more years. Then

Pearl Harbor on December 7 and America went to war. Japanese Americans were deemed part of the enemy, and 120,000 of them were ordered to evacuate, to leave their homes and be imprisoned behind barbed wire in desolate areas of the country. In August of 1942, the Masumotos were ordered to pack up what they could carry and depart for Gila River Relocation Center, south of Phoenix in the Arizona desert.

During the frantic weeks before evacuation, a question hung over every Japanese-American family: what would you leave behind? No one knew if and when they'd return. Some found a place to store goods, with a good neighbor if they had one. Homes were sought for personal objects and even pets — a dog or cat had to be disposed of quickly. Most sold what they could, along with the thousands of others in the area who were forced to conduct "fire sales," as Dad called them. "Pennies instead of dollars," Dad grumbles. Families desperately tried to dispose of what they could not take. Blankets, linens, household goods for a few cents. Small farm tools and equipment — good one day, virtually worthless the next. And Uncle George's new Chevy sedan, "sold for nothin'."

But the Masumotos had another problem. By mid-August, all Japanese were ordered to evacuate, and they'd leave behind a grape crop. Jiichan worried what would happen to the grapes that the family had pruned, tied, weeded, and watered; grapes that bloomed and grew fat through the spring and summer and hung ready for harvest in September; raisins to be dried and sold in the autumn, income finally collected once a year and profits split between the owner and renter.

The landlord widow was scared. She worried that without the Japanese to pick and dry the crop, who would take care of her land? Word had spread, and already in the plum fields of late July, workers could not be found, prompting both the Farm Bureau and the Fresno County Board of Supervisors to make an official request to delay evacuation, at least until the crops were in. The widow grew frantic.

She was too old to manage the vineyard for the last crucial month of harvest; she'd need help immediately. The best available farmers were busy with their own lands, and good renters were hard to find on short notice.

Dad approached, inquiring how she'd handle the farm. She said nothing. August began with no arrangement. A final irrigation water of the year was applied to the grapes, and they quickly swelled with juice. Meanwhile Jiichan made wooden crates from odd planks and old fruit lugs; he stenciled "No. 40551" on each box, the Masumoto identification number to be labeled on all personal baggage. The heat of late summer continued to beat on the countryside.

Finally Dad heard. The widow had found someone to take over that season's crop. The deal was this: the new renter would work a month and harvest the raisins. But he insisted on moving into the small farmhouse immediately. The tiny house the Masumotos had lived in for years "came with the rent" and had to be vacated in a day.

In two weeks, by the middle of August, all Japanese Americans would have to leave the area. But the widow demanded they immediately move. "She kicked us out," Dad said. "All she worried about was the damn grapes. Kicked us out so an Okie could move in." Dad mumbled and shook his head, "An Okie." The ugliness spilled from all.

They had a final meeting to settle on a price for the Masumoto year of labor. She gave them $25 per acre. The price of that year's crop would soar to $109 a ton or over $200 per acre. Jiichan and Baachan would get $25 an acre, and the widow and the new renter would split the rest. "No wonder the Okie wanted to move in. All he had to do was pick the grapes and count his profits for a month of work," Dad said.

"Now where do we go?" Baachan asked Dad.

The Masumotos were homeless. They had lost everything, cheated out of a year of labor, and in two weeks they would be exiled

to a desert with barracks and guard towers. Dad walk
road to a Japanese-American family friend, the Nakayar
had been renting a vineyard. Together they approached
kindly *hakujin* (white) farmer. There was a vacant house
more of a shack than a house. "Why sure, you could stay there for a
few weeks," was the response. The farmer would also give the families
a ride to the train station where they were to depart for the barren
lands of Arizona. He was a good neighbor.

As they moved out, Dad was furious. They couldn't take much
with them, a suitcase and crate, only what they could carry. His rage
burned. So it began – first with the few dishes. "The hell if I was
going to leave anything behind." Dad took them outside and
smashed them.

He grabbed pieces of the simple furniture they owned, some
built by Jiichan. He smashed a chair, and it splintered into small
pieces. He carried the wood outside and began a pile. He dragged a
table, a small stand, some other pieces, and flung them onto the pile.
They shattered and cracked into jagged pieces. He hauled out everything they had but could not take and tossed them onto the heap.

Then he set fire to it all. "Rather burn it all than leave it behind,"
he blurted. The wood snapped and hissed. Boxes ignited with their
contents melting. Papers and old letters and correspondence curled in
the heat. The flames danced in the late-summer evening sky; the
grapevines heavy with grapes stood as witness. The smell of smoke
filled the air, and Dad watched the fire engulf our belongings.

Now when I think of the scent of my father, I smell smoke in his
clothes. The aroma of his fire turns black.

Buddhist funerals often include a passage about "white ashes." "When
the winds of impermanence blow, our eyes are closed forever; and
when the last breath leaves us, our face loses its color. The body is then

sent into an open field and vanishes from this world with the smoke of cremation, leaving only white ashes.

"Nothing is more fragile and fleeting in this world than the life of man. Those who leave before us are as countless and as fragile as the drops of dew. Though in the morning we may have radiant heat, in the evening we may be white ashes."

Dusting Dad

HE ROLLS INTO the barnyard on the tractor, a cloud of dust behind him, late afternoon, another field completed. For a week he's been disking the vineyards for me, slicing weeds, smashing dirt clods, beating the parched earth in early August so that the soil will become a fine powder, preparing the land for raisins.

He ends each day with dust. Dust everywhere. Behind him, over him, below him, on him, in him, and with the right breeze in front of him. The disk, churning for days in dirt, has dust drifts that have collected on the metal frame, along every flat surface of angle iron; even on top of the narrow disk bearings a measurable amount gathers. The only parts of the disk that don't have thick piles of dust are the moving pieces. The tractor too is blanketed with layers and layers that collect along horizontal surfaces. When Marcy first came to the farm, I told her how we sometimes measured how long we had worked in the fields according to the amount of dust that collected. She'd run her fingers through the delicate pile that assembled on a tractor fender where the toolbox sat, grab a handful and let it trickle through her fingers, and say, "Well, I hate to tell you, but this isn't dust, looks like dirt now."

When Dad arrives, the smell of our dust accompanies him — a

hoary, desolate smell, a dry but not stale aroma. I know if I mix this with water, the earth will turn into a rich roe, yet for now it stands like sifted flour. I have trouble describing the scent, using words like "light," "parched," and "harsh," as if life has been sucked out of the rich humus earth. I can almost feel the delicate particles enter through and fill my lungs. The inner linings of my nose will be lined with a fine coat of the stuff, yet I know it carries much – pollens, molds, spores – because I'll sometimes sneeze as if to acknowledge the initial contact.

Dad pauses as he shuts the tractor off. His hat is tan, coated with dust. Any horizontal parts of him – his shoulders, his lap, even the tops of his hands as he holds the steering wheel – become depositories. Dad's white hair sticks out from beneath his oval straw work hat. The gray is masked with the dust, as if he tinted his hair with a cheap brownish dye which resulted in a peculiar coat of very unnatural light brown.

As he rises to dismount, dirt and dust tumble from these surfaces, creating a small cloud around him. He's a seventy-nine-year-old "Pig-Pen" from a *Peanuts* cartoon, lips tight, hair a bit unruly, streaks of dirt across his clothes and face, a constant billowing companion hovering around his ankles with each movement. He climbs down from the tractor with great care. He's slowed down in the last few years, his balance not quite as good as it once was, and bouncing on a tractor for three hours shakes you up quite a lot. The first steps – always a complicated climb down past gearshift levers and brake pedals and clutches, around various knobs and over the hump of the transmission that sits directly below the seat – you lose a sense of your legs for a moment, after hours of stationary sitting and bouncing. You almost need to prove to yourself that your feet are actually there. As Dad climbs down, I can see his dusty imprint still on the tractor seat cushion and a series of fingerprint impressions on the fender.

I delay talking with him, not to confuse with added verbal stimulation. He's an old man now and can handle only so many things at one time. A great piece of advice I too try to heed.

As his feet hit the ground, he turns to face me. His eyelashes have a veil of dust on them, and when he blinks, puffs dance in front of his eyes. "Man, Dad, you got a lot of dust!" I say the obvious, which is what a lot of farmers usually say in such situations.

Dad responds by nodding and saying, "Yep." Obviously.

I pat him on the back, sending another blast of dust airborne. My taps change to a slapping motion, hand flapping in one direction with the back of my hand returning with a whack. I move across his back and chest, following no particular pattern other than moving to a new area after two or three rounds and a pronounced reduction of dust. Initially Dad dusts off his sleeves, but as I keep knocking and stroking, he stands still. I keep hammering at certain surfaces heavy with dust. His shoulders are heavily embedded. I squat and work his legs, his thighs heavy with dirt, and it seems as if a few pounds are shaken loose, gathering around his shoes.

Dad's rather proud of the dust. He's worked a full day, the elderly farmer proving his worth. He's still valued. The smell of his dust is old but not obsolete. As I work my way back upward, over the back of his legs, dancing across his lower back, and revisiting his shoulders again, on his face an impish grin grows.

PART FIVE

The Art of Touch

A Farm Woman's Hands

Hands that hold.

Hands that plant, prune, clip, thin, lift, toss, haul, shovel, weed, hoe, pick, roll, unroll, and harvest. Hands that drive, steer, shift, hoist, turn, twist, and support. Steady hands.

Hands that sign, cosign, gather, collect, phone, mail, and calculate. Hands that cook, bake, mend, sew, create, craft, save, and fix.

Hands that bleed, crack, bruise, break, and heal. Hand labor. Hands that should have more time. Hands that are powerful. Hands that grow callused. Hands that can be soft. Hands that love.

Hands farms need. Farm hands. A farm woman's hands.

Hard Sweat

FARM WORK makes me sweat. It first appears on my chest, seeping through my pores, subtle and silent, unannounced on bare skin. A damp chill tickles, yet my body sends me a simple message – "hot." I increase my work pace and the sweat drips onto my back, then arms and neck. A wet patch spreads across my shoulders, announcing itself. I can feel my T-shirt and work shirt stick to my flesh, clinging to my body. I grow conscious of my heavy breathing. I can feel moisture in my hair. A drop dribbles down my temples. My forehead beads. I wipe away the perspiration, blink to keep it out of my eyes. I'm losing liquids. At this pace I can't go forever, have to stop sometime. Sweat reminds me of limits.

Farming is supposed to be physical. Manual labor is part of the process, a requirement to stay on this land, a necessary job. Honest work. I can't think of a farmer who doesn't do a lot of sweating. I have no love for the tedious, backbreaking work. I've tried to find easier ways for the hardest jobs. I rely on tractors, modify old tools, adapt machinery. If necessary – usually years later when I can justify the expense and my body remembers the work and hurt – I buy equipment. The ultimate solution, of course, is not to farm. (Gardening is the same – you will sweat and the only way to avoid it is to hire a gardener.)

Some jobs demand sweat. The best tool will never dispose of all the weeds. No machine can pick ripe, delicate peaches, and equipment seems to wait until the hottest time of the year to break down, out in the fields, farthest from the barn, late in the day, when I'm beat. A farmer who frequents the answer "No sweat" is either lying or fooling himself.

Even in the cold, foggy weather of winter, with temperatures in the twenties and thirties, I sweat with certain jobs. Broken grape stakes must be replaced during this dormancy. I pound new metal ones into the earth, slicing through rotted old redwood stakes, bound by vine roots. As I work, I peel off layers of clothes. A jacket marks where I started, a sweater at the place I felt warm, even a flannel shirt draped over a stake like a limp flag signals where I began to sweat. A chill dances across my chest, my body heat contrasting with the cold, creating steam when the two fronts clash. A shiver stops my work.

Summer work pushes me out into the hundred-degree heat of the Central Valley of California. Our heat is supposedly "dry" with little humidity. It helps. A dry 105 is better than a muggy 95. Still hellishly hot. I begin to sweat just by walking outside. Ubiquitous heat like a furnace with no escape exit door. Fool to be out in this kind of heat. But grapes and peaches demand it, and if I'm thirsty, so are they. Six times in late spring and early summer, I flood-irrigate, cutting furrows in the earth across eighty acres and spreading water down each row. Water is supposed to follow a path, reach the end so I can start the next set of rows. Never works that way. Breaks through dirt, cuts across a furrow, follows gopher holes, and skips rows. Under the scorching sun I trudge down, dam the break, plug the hole, right the path, and wait; come back an hour later and find another gap.

I come in at midday, peel off soaked shirt, hang it over rail to dry while wiggling into clean one. My treat for the day; fresh clothes. I can measure a day's work by the number of wet T-shirts draped over the porch rail. My wife, Marcy, comes home and knows how hard a day I had just with a glance.

Return to fields, give water to thirsty plants while I lose fluids. Lean over shovel, arms ache from moving dirt, back stiff from fighting water, sweat drips and splashes in the irrigation water. My leather work gloves are wet and stain the wooden handle. Moisture soaks the band of my work hat; the sweat will stain the straw, leaving behind a white deposit of salt.

I begin to labor. Sweat wanders down my face, always worst with the eyes. A salty burn and I'm forced to stop. I blink and cry, trying to wash away the sting. Wipe my face with my stained sleeve, mixture of moisture with dirt and dust, smear the sweat around on my face.

Some years tropical moisture swirls up from Baja California and couples with a high-pressure heat wave for a combination blast of humidity and 105 degrees. Hurts to breathe. Sweat so much only the belt loops of my pants are dry. Perspiration soaks through socks and stains work boots, penetrates my leather belt, work shirt clings to my flesh, my straw hat droops with moisture. During a break I come inside to hide. Strip down to nothing and stand still, still sweating as if I had just come out of a shower. Set everything out in sun to dry; the clothes bake in minutes, curing stiff with a dingy yellow salt stain in middle of the T-shirt, along pants waistband, and across my work hatband.

The work and heat beat me up. I lean over, panting. Take short breaths, saving energy, feeling the heart race, and veins and arteries pulsating. I want to quit, not worth it, sweating is cutting years off my life. A drip off my chin, my nose, draining life away.

But more work needs to be done. Wooden props need to be staked under drooping branches laden with fat peaches. Dead limbs have to be hauled out of the fields. I drag them along the ground like corpses, casualties of war. Weeds grow wild, shading grapes, sapping moisture and food. And the peaches and grapes cry out for more water. And hard work. And hard sweat.

Is this what I must do to keep farming – do more, work faster and more efficiently? Is this how the struggling craftsman or starving artist must work?

An old farmer told me sweating will harden you. I thought he meant I'd learn tough lessons about life. I know better now. Hard sweat drains me of youth. Now I feel old, years lost with the hard, physical work of farming. I've changed quickly from a young farmer to an old one with nothing in between. Sacrificed years for a juicy peach or sweet raisin. The work beats me again and again. You may start out a young farmer, but after just a few years of sweating, you discover you're old.

I stagger home and collapse on a bench on the front porch. My children see me, they worry, wonder if something is wrong. "You okay, Dad?"

My eyes stay shut. I shake my head. No energy to respond. "This farm's gonna kill me."

They stand silently. They know I'm telling the truth.

Stay Out of
the Fields

Two years after our daughter, Nikiko, was born, she wandered into a vineyard I had just sprayed with an insecticide to kill vine hoppers. These pests numbered in the tens of thousands, sucking at the leaves, draining nutrients and life. The upper leaf canopies had already withered brown from the insects' feeding. I applied a poison, and within hours the hoppers would quickly tumble onto the ground, killed efficiently.

Nikiko was learning a new skill of walking, and driven by a healthy curiosity, she quickly slipped out of my sight. I grabbed her before she fondled the wet leaves. She cried as I dragged her into the house. She could have been poisoned. She would have gotten ill. Instead of setting her on the couch, I tossed her roughly. "Don't ever go into those fields!" I towered over her.

A look of horror struck her face. Her lips trembled and her eyes grew glassy with a frozen, piercing stare of disbelief and confusion. I was frantic, but caught myself and stopped. We were both terrified. Something was wrong, very wrong. This all seemed to contradict the meaning of a family farm.

Once in the middle of summer, Dad broke out with an odd rash. His neck and arms grew red and inflamed, and a collection of welts grew and festered. The stoic farmer who rarely complained did this time. Wherever his skin was exposed to the sun and mixed with sweat, the flesh was puffy, swollen, and burned. Even his eyelids began to swell.

I asked where he was working. In the vineyards, mostly on a tractor. I was perplexed – I had recently sprayed the field, but the official reentry period was long past. It was supposed to be safe. We tried to solve the problem with a fresh set of work clothes while Mom washed his old ones in hot, hot water. We also applied a variety of creams and ointments. After another day outside, the pox spread deeper.

Dad couldn't sleep at night and finally consented to a doctor's visit. Medically, they couldn't pinpoint the problem. He was given a shot, antihistamine, and told to stay out of the fields.

Stay out of the fields? You simply don't tell a farmer to stay out of the fields and expect compliance.

But something was wrong. Was it something I had sprayed? Nothing I had done was illegal, of course. Most of our treatments contained the least toxic chemicals registered for use on our farm. Yet I was unsure, and the doctor's words haunted me. Staying out of the fields would emotionally kill my father. I knew he wouldn't obey, nor would I.

I attempted a compromise with Dad. "Okay, work, but stay on the tractor. Don't go touching the grapevines." At first he followed my instructions, then I spied him climbing off in order to reposition a loose vine cane. He couldn't help himself. The rash spread to wherever he was perspiring. His temperament grew short, and he snapped at everyone.

Mom pleaded, "Stay out of the fields!" He didn't answer.

I had to make a change. A son isn't supposed to hurt his father. The place belonged to family first.

So I now farm organically. I stopped using toxic chemicals, and utilized integrated pest management — good bugs eating bad bugs. I use compost and organic fertilizers and grow cover crops — clovers, legumes, and native grasses (neighbors still call them weeds) — in order to increase organic matter in the soil, which provides a habitat for insects and helps with water penetration. I try to work with nature.

But organic farming is slow. I can't apply a quick fix and spray to kill a bug or spread a nutrient that is rapidly absorbed to change yellow leaves to healthy green. I can't simply fix and repair by adding large inputs of capital or technology. I had to learn an alternative approach and a different philosophy in order to get in touch with my farm — not controlling but cooperating with nature. This required time, fostering a new relationship, courting a new perspective, formulating a revised vision. Management replaces other inputs; more and more of my time is spent listening to and keeping in touch with the farm. I walk more than ever, I have a physical relationship with the land. I no longer separate the places I live in and work with. It's slow work, and that's fine.

I seek methods that take advantage of my small-scale operation. Often, I rely on simple tools — an old pruning shear with a curved blade manufactured locally, perfected over the years by complaints and suggestions from the vocal local farmers. A shovel has become my favorite and most reliable companion, replacing herbicide sprayers and large plows. I use an old worn disk with blades that only scratch the surface where weeds lie but gently pass over the hidden world of microbial life that thrives deeper below the surface in a healthy soil. With such tools, eighty acres of vines and trees seem about the right area for our family to work, a proper ratio of eyes-to-acres and hands-to-lands.

I hoped that organic farming would prove not only environmentally sound but also economically viable as I searched for a market

l organic outlets could help differentiate my old
re not a generic commodity like fast food. I believed
ɔposed to be both grown and eaten with care —
d story.

Yet I still have nightmares of years of my own life stripped away
by the hard, physical work. I worry that spending the extra hours and
hours of time in the fields takes me away from my family. The work is
not getting easier. Contradictions continue to live in the shadows of
my farm, looming in the dark under grape-leaf canopies. Working
slow with long-term investments of time remains a huge risk; I
blindly step forward, compelled by optimism and naiveté.

My vines and trees require patience, because their change is
gradual. My old orchard has a character composed of many variables:
weather, age, fertility, water, and soil life are just a few. When I first
returned to the family farm, part of me wanted to organize work
plans, schedule irrigations, calendar pest-control treatments, and
remove as much variability as possible — all for the sake of efficiency.
But farming is not done by formula, a lesson I quickly learned when
the spring rains did not cooperate with my intention to manage weeds
by a staggered schedule of monthly disking. The weeds grew despite
my intentions.

Now I try to respond to the elements. I disk according to the
intervals between rains, and measure my work in terms of long timelines. It will take a lifetime to get to know this place. I learn the stories of these fields only as they unfold naturally and gradually. I expect
the process to be complicated, mixed with the right amount of chaos.
I don't want simple flavors, those that result from following pre-
scribed farming formulas, in my peaches and grapes. I hope for flavors
that can be wonderfully complex.

I enjoy this challenge. It requires the creativity to fix and sustain.
I think of my work as larger than myself. A farm is owned by generations; an old peach orchard or vineyard is created by the legacy of

many. The power of stories connects me to this place, bonds me with the land, and envelops me within a rich history. On my farm, the old stands along side the new, and daily I am touched by a continuity with the past. Harvests come gradually and are savored slowly. This is how and why I farm.

Mom's Hands

I FORGIVE MY mother for being different. As I grew up in the 1950s and early 1960s, my mother's hands grew hard and rough from working in the fields. I sometimes wished she'd work inside, perfecting culinary and homemaking skills. But her hands could prune vines better than make a roast, and she could pick and pack peaches with the speed and the artistry that could have been dedicated to perfecting pie crust and fruit tart recipes. On the farm, we valued her for her callused hands.

In the winter she tied grapevine canes to the trellis. She wore cotton gloves with the fingertips cut off, partial protection from the cold and brittle wood, the material helping her grip the thick branches that needed to be gently bent and coaxed into a horizontal position and wrapped around and around a wire. Most often the ends of the canes were braided into each other, held in place by slipping the end into a gap of another wound limb. Sometimes a "quik-tie," a short, thin strand of wire embedded in a paper strip, was used to fasten and hold down the temperamental canes. Exposed fingertips of one hand held the vine while the other braided or lashed it in place. Our winter temperatures usually only dipped into the low thirties, but our foggy and overcast days kept humidity high, a wet cold penetrating into joints and exposed skin. Stiff grips often gave way, a cane slipped loose and slapped a rosy cheek, shooting pain into the flesh, and later the tears rolled down faces to soothe beet-red welts.

During summer harvests, Mom stood in the number one packing stand, a station at the head of the sorting table where the fruit rolled off a conveyor belt and onto a flat area for inspection. Then she hand-packed each fruit into boxes. I recall her snatching a peach and rotating it in her hand, feeling for defects, searching for a soft, overripe spot while judging size by the shape and weight. Her glances darted from the pile of fruit back to the box in front of her as her hands grabbed and squeezed. Finally her gaze was pulled to the fruit in her hand for a visual second opinion. Slowly she packed the fruit in a pattern so that twenty-four peaches fit in a single layer, then a thin cardboard shim and a second layer. Other times I watched her ponder a choice of whether to toss or keep a fruit, and with an honest touch, she'd grudgingly toss the cull, her touch overruling her eyes. Bad fruit was tossed into a cull box, good fruit snugly packed in the box, hands in constant motion, searching for the right size to finish a layer, flicking an obvious defect. Although both hands worked independently, I detected a pattern: the right sorted more, and often good fruit was handed off to the left hand, which gently nestled the piece into a home.

Later, when I was still in elementary school, Mom's hands left the fields for an office. She worked filing and typing, part-time at first, then full-time with her vacations reserved for summer and our fruit harvests. Finances had become a problem and the farm needed a hand. The "city job" provided not only additional income but also stability. Hands could be kept busy and be paid.

I remember thinking we were no longer a family farm, worrying that the need for off-farm income was a sign of failure for Dad. Mom became the breadwinner because we couldn't make a go of it from the land alone? Then I thought of a children's story about a farmer and his family who annually took goods to the city and bartered the family's products: goods from the farm, like wool and hides and also the hand-crafted furniture, wooden toys and knitted mittens. Farm hands often generated off-farm income. Perhaps Mom's work wasn't that different; historically, farm families worldwide always supplemented their work. My view of the American farmer was fabricated from a 1950s myth based on an "agribusiness" model of specialized production and narrowly defined roles and job titles.

But Mom felt bad because she didn't like leaving me alone, a farm "latch-key kid." I insisted I didn't qualify, because first, we didn't lock our farmhouse, and second, Dad knew my schedule, and even from the other side of the ranch, he could see a bus dropping me off and a small black-haired kid bouncing down the country lane under a pale blue sky, surrounded by fields that I knew better than most anyone else. I believe that image of Dad's watchful gaze comforted Mom. During her office years, her hands grew a little softer, but a weekend of pruning or thinning trees or the after-work long autumn evenings of boxing raisins kept them in farm shape. Most of the calluses stayed, and she never lost her farm touch as a measure of who she was and is.

When Mom turned fifty, she began having hearing difficulties, and within ten years she lost her hearing completely. Friends often asked me about her, offering sympathy by saying, "How sad. It must be hard for her!" I'd sometimes quip, "It's not as if she died."

My mom gradually became different from her peers and lived in a silent world. She could have easily withdrawn, her kind nature slipping into a lonely world, her Japanese upbringing telling her to be quiet and invisible. But she learned to adjust. In her own way, she became assertive by letting others know she had a hearing problem, by jotting a note down and inviting others to write her a message. By her doing so, a burden was lifted; she stopped blaming herself. Others would have to find alternative ways to communicate with her. Unfortunately, too often people would ignore her message and instead shrug and mumble that they didn't know sign language, kneading their hands together, paralyzed as if they too were disabled.

My mom learned to cope and live with her loss. "At least I still have my other senses," she'd say, and she began writing all sorts of notes to herself. She snapped her fingers regularly with a new idea or revelation, or sometimes when she recalled a thought during a senior moment.

Recently after years of feeling dependent upon Dad and the family, Mom seems to have gained a new sense of independence. At seventy-three she takes charge of her projects. When she decided to plant a vegetable garden, she didn't ask us to help. (On a farm with eighty acres of peaches and grapes, sometimes a garden is the last thing we want to worry about.) Instead, she asked us *how* we'd like to help. Everyone contributed with the shoveling and weeding and picking, although she did most of the chores. As if to ensure her vegetables remained hand-grown, she conspired with Dad to make a fence around her plot that's too narrow for me to back in a tractor. All the labor is still done by hand. Whether we wanted it or not, we now have a family garden.

The most recent project was when she grew a peach in a small jar with a narrow neck and opening – the farm woman's version of a ship in a bottle. She had seen it in a magazine with some other fruit, and, inspired, she found a small round juice bottle about five inches in diameter, short, with a skinny neck less than an inch wide. Just after bloom when the first peaches set, a tiny little teardrop-shaped fruit and its branch were gently bent into the opening of a collection jar. She started with about a half-dozen, knowing some would not survive.

Over the next few months, she monitored the progress, nudging the growing fruit farther into the bottle, adding a complex structure of strings and tape to keep the glass enclosure at the proper angle and supported on the tree branch. A few jars slipped off during a spring wind, the fruit withered in others, but a couple actually grew and grew. By harvest time, a fat peach swelled within the glass enclosure, coaxed by Mom's hands, nurtured by her now daily massage. Finally she plucked off the peach, filled the container with vodka as a preservative, and set it out to show friends and visitors. They couldn't believe it when they saw it, even after the story was told, thinking Mom used some type of trickery. "Just my hands," she said, happily unable to hear their doubts.

I hope in the future, my hands become like hers — each callus carrying a piece of our family's history. Every spring when the peaches blossom, I think of her as I walk in the translucent light of a field in bloom, the vibrant hues creating a rich glow with a luster that envelops. I pinch some of the petals, and they feel like the wings of a butterfly. Quickly I let go, thinking I've damaged the delicate tissue. Within days, though, the fragile flowers tumble to the earth, and I'll think of Mom's peach in a bottle, remembering just how tough nature can be and the differences Mom's hands have made. She made our family blossom.

Touch as
an Artist

B AACHAN/GRANDMA lived with us on and off throughout her life. For the first twenty years of my life, I knew very little Japanese and she used no English, yet she spoke to us in many ways. Years ago when we were children, she'd massage our backs, her rough fingertips running over the youthful skin, each callus generating a natural friction that warmed my soul and filled a memory. Now I realize her dry, cracking hands actually rubbed off the top layers of our skin, exposing a deeper layer of touch receptors. She made us more sensitive. I also have thought about each callus and how it came to be — only through decades of hard work in the fields. No one happens to have calluses, they are earned. Her touch helps me complete her picture even years after her death. I can still feel her kneading my skin, which triggers other senses.

I can smell her old aroma. Clothes that have been sitting in a dresser too long, that have been worn too many times between washing, a stale fragrance at first offensive, then gradually oddly comfortable, because you associate it with someone close. Her breath, not rank nor foul but musty, like an old pair of shoes you still keep around.

Remembering her touch, though, triggers the sound of her voice, in a foreign language for most of my life, alien to this land, lost in a distant place she may never have called home despite living in America for sixty years. A deep voice for such a small woman. Under five feet tall, she lacked the dainty, delicate tone I'd imagine of a petite Japanese woman. She came from the working class, and her sound matched the coarseness of her massage – simple and firm.

Baachan's hands cut through generations, a touch of a grand-mother and grandson and a story of a working woman on the land. The more I learn of Japanese Americans – immigration, relocation, and agrarian history – the more I can feel the pressure points in her life with each hardened callus.

Touch beyond hot and cold or pain and pressure. Marcy tells of a time when the lack of touch pierced deeply. She grew up on a goat dairy. Her folks claim Marcy's allergies to cow's milk made them venture into such a marginal endeavor. Goats – before their cheese was gourmet enough for the general population to demand, before health food stores expanded with new clientele clamoring for alternatives, back when goat people suffered from a poor image compared to the beef and cow folk of large, established ranches. Goats, the stepchild to the dairy industry.

Marcy's family worked hard, tried to forge new markets, and struggled. "We were before our time," claims Florence, Marcy's mom, with a sigh and chuckle.

But like good ranch families, in addition to their mainstay of goats, Marcy raised sheep and beef for 4-H. She'd take a young calf, nurture its growth for just under a year, fatten it up for the county fair. "You can't help but develop a relationship," she claims, "even though you know what's inevitable." A few, with their big black eyes, became favorites, but she rarely thinks of them, because at the end of the year, her 4-H project completed, she had to say goodbye.

At the fair, she and hundreds of other kids had spent hours and hours with their animals, caring and preparing for showing. Some even slept on straw beds and sleeping bags with their animals in their assigned stalls. Then near the end of the week, she'd groom her companion for the last time, one final feeding, watering, and curry brushing. Leading with a halter, she'd parade a different one each year — Sarah or Buster or Big Bob or Mae West — for a final walk through the show ring. (Marcy didn't have the twisted humor of others who'd use names like T-Bone or Meatloaf.) As they walked, both her name and the animal's name were called out, along with awards they had won as a team. And of course the weight.

The auctioneer would begin his chant, shouting a starting price in a rhythm, warming up the audience. An arm would shift, a hand would lift a paper card with a number raised by a bidder, and the dance would commence. With luck, the duo of seller and buyer would become a trio with an added buyer, or better yet a quartet or quintet, competitors bidding against each other. More often, though, within a few seconds a final price would be matched with a buyer, and with a quick tap of the gavel, the pet would become someone else's property.

The next day, a truck or trailer would pull into a loading area, and with a nudge, Marcy would lead her buddy for their final walk together. Unbuckling the straps, in a single, smooth motion, the harness would be removed and the animal shooed up a ramp. "The hardest part," Marcy explains, "was standing alone, fighting back the tears. Empty harness in hand." Sometimes the greatest emotion is when you're no longer touching, no longer holding on.

I try to map the touch world on my farm. Pockets of the different soil types are charted first. Using topography, I outline the small island of clay on the western edge of a vineyard, the finger of sand running by my folks' house and the hardpan plateau sitting just below the sandy loam in my Spring Lady peaches. I've gotten to know each type in different seasons. With late-winter rains, the clay holds water and I can sink deep in the mud; if at all possible I avoid the area and warn tractors to stay away. Sand is forgiving, and I can enter those fields soon after a storm, cheating on my work schedule, knowing most weather can't keep me out. The hardpan with its thin skin of topsoil will never allow deep roots to penetrate; the trees will forever suffer no matter how much I add to build the earth, lifeless and coarse.

I can adjust my map to a micro level when I think of the coarseness of old tree bark or the gnarled trucks of hundred-year-old grapevines. A pest to stone fruit, San Jose scale, hides in the cracks and rough bark during winter. I can coat the bark with a fine mist of dormant-season oil, which will suffocate the insects, provided I cast sufficient coverage. My sprayer has to envelop the tree with a shower to the point that it makes the smooth young wood shiny and creates a trickle that will flow down into the crevices and fissures of thirty-year-old bark. I have to imagine this happening as I drive slowly, very slowly, down each row, my tractor pulling a five-hundred-gallon sprayer, drenching each tree with twenty-five gallons at a time. I'm impatient to reach the end, anxious to emerge out of the fog that's soaking me with even the slightest gentle shift in wind. Otherwise I will drive too fast – perhaps a sprint at three miles per hour – and the spray will fail to blanket the surface, sparing the sucking pest.

My farm work can be scaled according to touch. Certain shovels work for specific jobs – a rounded blade for trenching with powerful strokes, a flattened cutting edge for surface weeding with a smooth, gliding motion just below the surface. For tractor work, I have modified a disk and attached it to a hydraulic arm. I can scrape weeds away from a tree trunk by feeling the blade rub against the wood as I pull and tear the knotted weed clumps away. I need to take a gentler swipe when working with young trees and their softer, delicate bark, but the ancient Sun Crest orchard, like an old bear waking from hibernation, wants to be scratched as weeds are yanked from around its roots. It all works best if I'm able to imagine the machine as an extension of myself, the blade like my hands searching for a grip to rip the balled roots, seeking a finger hold to fasten itself. Then as the machine jerks, I want to feel the sharp disk blade edge grating against the bark like fingernails scratching a chalkboard. I grimace but know I am not gouging the woody membrane. Different techniques work in the various orchards of different ages.

I often wonder if fruit shouldn't be marketed via touch instead of cosmetic appearance. What if my peach display was accompanied by a single soft, gushy one right up front, perhaps on its own pedestal? We'd sacrifice sale of that one, because by the end of the day, the bruises would show and juices would be pooled in a sticky puddle. But shoppers could feel and judge the ripeness for themselves, fulfilling that irresistible urge to squeeze I've seen in both adults and children — like the popping of the air pockets in plastic bubble wrapping with a series of snaps and pops. Rarely can they just stop at one.

The audience for these peaches can then leave their mark on the display fruit, getting their daily tactile fix by touching something soft, warm, and handmade. The sensory contact may generate a higher level of satisfaction, a stimulation of all the senses. Then, in my wild dream, people take home my peaches and secretly kiss them — a slow, lingering kiss, engulfed with not only a sense of touch but also taste and smell. A deep breath, lips part. Tongues replace the fingers and run along the surface. Feeling, not for imperfections, but just feeling. Then teeth gently press into the flesh and a complex touch dance begins.

To support weak branches, we use hundreds of wooden props, long, skinny planks ranging from eight to ten feet in length and only about one inch thick and two inches wide. A piece of heavy wire is curved into a crescent shape and loops around the wood near one end; this "prop hook" can be slid along the length, adjusting to the random height of a sagging branch or limb. As my pickup bounces down our dusty avenues, a hundred props rattle and shake, threatening to jump out. Transporting them from my barnyard, a half mile to the folks' orchard, I keep an eye on my rearview mirror, checking to see if any escape. The fine dust of years of service puffs into the air from the stacks with each bounce.

Dad waits in the shadows along the edge row of the peaches. Behind him, high in the trees, soaking up the long early-morning direct sunlight, the fruit has begun to grow fat, and against the dense green of the canopy of leaves, the deep red of the peaches makes them look like velvet globes. Dad excitedly points to the interior of the orchard and beckons me in. We trudge to a section about five or six trees from the border where a series of limbs bend low to the ground. None are touching the dirt; most lean down to eye level. We're still in time for rescue. The heavy weight of the fruit pulls the branches down, and the scaffolds look more like a weeping willow than the rigid V-shaped and properly trained orchard tree.

These are vigorous young trees, about six or seven years old, and their major scaffolds, only three or four inches thick, tend to bend rather than break. The bowing damages the wood, straining the fibers and stretching the cambium layer, weakening tissue forever. Once a limb has drooped too far, it will not bounce back up after the weight of the fruit is removed. Next year we will try to correct the balance, training the wood upright by tightly wrapping rope around the top third of the tree. But where the bark was severely twisted, stretch marks remain and can never be repaired. The limb will forever sag outward, even pulling the rest of the tree slightly to one side, creating an unbalanced structure. If the branch is on the east side of the tree, the afternoon sunlight will penetrate the interior from the west, and the intense rays from overhead will sunburn the bark, scorching the delicate surface. The skin will dry, curl, and peel, weakening the branch more and creating nooks and crannies for disease and pests to invade. During the winter, even the best pruners cannot reshape the damaged tree into the proper vase form.

Dad wants to prop the drooping branches, returning them to their original position and supporting other limbs before they bend or possibly break. I remember as a child following him into the orchard. I threw dirt clods at imaginary monsters while Dad carried a load of props on his shoulder. He'd prop in the evening after the long, hot hundred-degree daytime heat had peaked, during the summer days when I could play and he could work past eight o'clock, even until nine. He'd hoist a dozen of the wood stakes and march from tree to tree, shoving a brace under certain branches, anticipating problems before any fruit was lost or wood damaged. He called it "saving a branch," since we had "brought the crop this far."

Once, a rare late-July thunderstorm struck our valley, bringing rain and wind. The trees shook with the gale and fruit battered against branches, tossing leaves with every movement. Lightning flashed, and thunder pounded the heavens. But with one blast of wind, a shot pierced the air, roaring above the storm, the distinct crack of wood shattering. Next to our house, a nectarine tree had exploded, the rope bursting, the limbs crashing, the few props stationed for support crumpled under the sudden weight and velocity. We thought for a moment lightning had struck, yet there was no flash, only the crushing sound of a tree collapsing. All of the main limbs had splintered, and the tree looked as if it had been slapped and beaten, knocked down, bent over, limp and broken.

Quickly the storm passed, and we all slipped out to access and try to repair the damage. A twelve-foot-high tree had been severed in half, limbs swaying in the breeze, fruit scattered across pools of water and hail. I concluded that one major branch had given way, and that the others followed, unable to stand alone, trapped in the howl of the wind. We couldn't do much, so we left the drooping branches to find out if any had enough wood still attached to feed the fruit a few more days, though we all felt the impulse to bury the dead.

When I was a teenager, we propped every year and just about every tree, rescuing branches with a heavy crop of fat fruit, saving precariously leaning trees that would snap under their own weight. We rotated hundreds of sticks from the first peaches to ripen in June, then hauled and set them in nectarines and plums, back to midseason peaches, ending with August nectarines. Gradually we propped less and less, wrapping ropes around the tops of the trees to hold the branches in place and keep limbs from drooping or sagging. Now I try to train the trees into a tighter vaselike shape, hoping to avoid the extra labor and time of props. The rough prop wood has weathered over the years, and even after decades, fine splinters often break off into my hands when I carry them.

When I first returned to the farm after college, I wagered a few branches were acceptable casualties of the battle toward harvest. I wanted to leave as much fruit on the limbs as possible, gaining an extra box or two from each tree, pushing the limbs to their maximum. I expected to have a few break every year, then I'd learn the limit of these trees.

But Dad was from another era when we couldn't afford to lose any fruit or limbs. Every summer, he retraced his steps from years before and wandered from tree to tree, propping branches and limbs, often the same branches weakened in their youth, improperly pruned and trained or simply those that leaned a lot. I'd walk with him for a while, questioning why we'd spend so much time on the decrepit that would soon be lost anyway. "Euthanasia," I joked, a word he didn't understand. Eventually he'd outlast me, carrying armful after armful, walking, searching, propping, saving. I gave up and started other work, swinging by later in the evening, listening to him drop a load of props in the middle of the orchard. The wood sticks rattled with a muted thud. Then I could hear him sifting through the small pile looking for a certain length to match a nearby limb. Dad's propping walks were not just about maximizing monetary profits. He cared about these trees, knowing well a broken limb would take years of training new wood in order to replace the old. The pile of fruit rotting on the ground beneath a severed branch would appear as public reminders of failure. I think he felt shame with each broken limb.

Dad likes to prop in the early morning or late evening. Early to catch a limb leaning too far, late to rescue one before nightfall. Older now, Dad stumbles through the uneven field when carrying an armful of the long wooden sticks. He angles one midway up a thick branch, and I'm reminded of the Japanese character *hito*, which means "person."

"*Hito*, written in two strokes," a farmer once told me, "a long one with the other holdin' it up." The old farmer's hand mimics a long, even diagonal stroke, followed by another short stroke midway and at a right angle downward, as if it holds up the first. "A farmer can't stand alone, has to lean against someone."

Dad tends to his old friends, revisiting them each year. They lean against each other. I think he also enjoys propping the younger branches – they too need an old man to care for them.

Most large farm operators avoid propping. I believe they find the step too costly and time-consuming. Farm managers don't like the unruly growth and take a chain saw to tame the misshapen trees. I wonder if good trees don't naturally age and begin to lean, branches pushing outward in their search for sunlight. I've sometimes theorized the best peaches come from limbs that gradually bend during the course of a summer, dropping inch by inch and opening their interior canopies to life-giving light. Perspectives change, trees develop a natural curve, and the sun can penetrate deeply while peaches bask at slightly different angles each week. Part of the accidental art of a glorious-tasting peach?

That doesn't mean I love to prop the way Dad does. I understand more now, but it still takes time from other work, and the props often prevent a tractor from entering certain rows. I watch Dad at work propping. He has a content look as he gazes up, eyes searching for a friend in need. I wonder if he's trying to cheat death by rescuing a few limbs. Or is he simply trying to fix something before it breaks? I'll miss his touch.

There is an art to propping. It takes time and a touch for creating just the right angle. The rhythm is slow and tedious. With an armful of four or five props, I have to pause to study a tree, walk around it, looking for sagging branches or leaning limbs. Summer has moved into the field; I sweat in ninety-degree mornings, dangerously hot afternoons, and long evenings with stagnant air. Orchard floors are weedy or filled with lush cover crops, pleasant to the eye but terrible to trudge through as I march from tree to tree. I have to wear long sleeves as I cradle the sticks – the most efficient method to carry a bundle – because I'd rather they leave their dozens of slivers in my cotton shirt than flesh.

The Sun Crest orchard remains the hardest to prop. Some old branches are ten inches thick and appear strong, a testament to their three decades of growth. In the winter, I rope every tree with a single or sometimes double strand wrapped around the tops. The cord helps all the branches support each other; a sagging limb on the south side is sustained by the northern ones. But some angle outward, a limb growing precariously away from the base, creating a wide V and a weak angle. In other cases, young new growth has squeezed in between these ancient branches, the height not quite high enough for the tree rope to catch and support, yet the limb is too fat with fruit to sustain its own weight. As I walk I realize I can't prop every limb – five to six thousand supports would be needed per acre. So I guess which need help the most.

The correct prop must be placed at an angle to support the

weight. My eyes initially deceive me. At first glance, it appears that all that matters is to shove the prop hook under a bearing branch, push it upward as far as I can, and plant the base of the stick into the dirt. I've tried this method and found myself trying to balance the weight as it shifts to and fro, only to secure the limb with a second prop in triangulation. Two props to brace one limb does not make for good math and requires much too much effort. I now work by feel, trying to support the weight from the branch and allow it to rest on the prop. Good propping redirects the stress point and evenly distributes the pressure. We learn to lean on each other.

I will still lose some limbs each year, a causality of the season, a price for too much work in too short a time. Were this just a job, I wouldn't care. But like my father, I can't help but be troubled when I walk an orchard and notice in the distance an odd angle of a limb or a dark patch of green too near the ground. A branch is leaning and crying for help. I may then grab some props and rush to the scene, hoping the limb was just leaning too far. Sometimes I arrive and find the white interior bark cracked, fibers snapped, severed and jutting out like a compound fracture, bark dangling and exposed. "Damn," I whisper, and suddenly I'm acutely aware of my limits. I understand I can't save them all; that doesn't stop me from trying to test my limits.

I pause after propping a complicated tree – four or five thick, old branches ready to snap, and new, younger wood heavy with too much fruit for their young frames. I can see the rope strain and fibers fray from the weight. Seven props surround this ancient monument at different angles with varying geometric perspectives. Are my wooden stakes a stopgap for what will inevitably bow lower and lower with each year? Or do the lines represent something else, angles of mutual trust, a symbol of the human touch as I manipulate nature, coaxing another harvest from her? After a while, I no longer notice the props and instead see the tree like a bonsai, decades old, the branches impregnated with more sunlight, with the weight of a harvest.

Dances
with Weeds

FOR AN HOUR I had been driving through the sprawling city of Los Angeles, toward the hazy skyline of downtown. It was difficult to believe that in 1948, as my dad cleared hardpan rocks from the first land the Masumoto family owned in America (the rocks meant land prices were so low a poor farm family could afford a down payment), Los Angeles County was the largest agricultural producing region in the state and nation. Now, houses have gobbled up some of the world's best farmland, and freeways cover ancient truck farms of lush vegetables and strawberries.

I was heading toward the national conference of Dance USA, an umbrella organization for dance companies, dancers, and their supporters. Often ignored within even the artisan community, dance struggles for recognition and faces an uncertain future. I felt as if I had entered a hotel full of peach and raisin farmers.

Adjacent to the lobby stood a sleek pool of water, so clear it cast an illusion with seemingly no division between the bottom and sides. Along one edge the pool appeared virtually seamless; one moment you walked on tile, the next in shallow water. I dipped my shoe to see if it was really water and watched the ripples shimmer. My steps left a muddy footprint behind as I deposited a little of the farm in downtown L.A.

Carrying a shovel, I walked into the elevator and was greeted with a grin from a sophisticated, very blond older woman smartly dressed in bright red with a splashing black-and-white polka-dot scarf and matching bow in her hair. The smile seemed sincere, and I nodded my head.

"Garden?" she asked.

I shook my head. "No. Farm."

End of discussion. We both wore golden tans created from entirely different sources. Mine just up to my elbows and rolled work-shirt sleeves; I was positive she wore no tan lines.

As the elevator chimed I kept turning "garden" over and over in my mind, and I realized she may have been asking if I was the gardener or groundskeeper for the hotel. Thousands of Nisei gardeners once kept L.A. green for a generation until their Sansei kids made it through college. Perhaps I was a remnant of this past, promoted from suburban homes and lawns to downtown hotels and planter box landscaping. Keeping anything alive within these concrete canyons had to demand skill. I shifted my shovel to directly in front of me, leaning on it proudly.

In the lobby I saw a slender man in his thirties, smartly dressed in black. A former dancer and now an administrator with a small performing group, he asked his peer about the luncheon keynote speaker, whispering something about "farmer." She gave a sigh and glazed look. Even though they were artists, I felt my presence was going to push their creative envelopes.

The executive director of Dance USA had invited me, convincing me a fresh voice would be welcomed. Bonnie was well respected, but knew the risks of bringing a farmer to a dance conference. As lunch desserts were served and coffee poured, she wisely introduced me with a story about a conversation I once had with President Clinton about support of the NEA and NEH, a topic I'm sure the President did not expect during a meeting with California farmers in Fresno. Instantly my credibility with the dancers rose significantly.

I walked onstage with my shovel and announced, "I understand the uncertain nature of dance in America. I too have faced the harsh reality of defunding. My challenges arrive in the form of a hailstorm or September rain or March frost. Yet I have also made a thrilling discovery for the benefit of all craftsmen and artists: I have finally found a profession that makes even less money — farming!"

Immediately the audience relaxed into a burst of laughter and smiles.

"And I share an affinity with many of the organizations here today. Though it was not my intent, I have now joined the ranks of nonprofits." A few applauded and chuckled.

"Dance and farming share a physical relationship with the world. Movement dominates our daily rhythms. We translate our emotions of caring into a cadence, a beat. A simple shovel becomes a dance partner. Daily I engage in a routine — dig, pat, pound, toss, fling, hurdle, poke, stab, retreat. Each motion has its function and is carried out in concert with a natural tempo."

I dropped the shovel head and scraped the blade along the stage floor, swinging and lifting, pushing and retracting. I described the work of spring, reenacting the dance with moist earth that welcomes the steel edge as it severs roots and glides through the soft texture. With the hoary dirt of summer, drained of moisture, I change the angle of my cutting, skimming just below the hard surface and flicking weeds to the side, soon to wither a lifeless brown. Then with the winter's cold, sometimes frozen hard, a dormant season for repairs and dreams as I punch and dig holes, planting bare roots of hope. This was my *Rodeo*, combining folk dance with the ballet of nature. Agnes de Mille would have liked this.

"I dance with my shovel and now know this: at the end of each song and each season, I still have hope. Our dance is a labor of love. Without hope you wouldn't go on to the next year. We give ourselves to the world, and I consider it a gift, because while there are other easier ways to make a living, we choose to do what we're doing. No one demands that I farm, just as no one forces you to dance. We must live in a world of optimism. Blind optimism at times."

I then surprised the audience by asking: "Despite our struggles, who considers themselves wildly optimistic?" About thirty raised their hands, and I insisted they stand. "You have faith in a complex world. We share a naiveté, an unyielding sense of hope. And" – I paused – "we may be possibly the dumbest creatures on earth." I then honored the brave souls with a gift of peach jam Marcy and the family had made, wanting them to take home the flavor of a farmer's optimism.

I told a story about once failing with our family garden and the squash and tomatoes my then eight-year-old daughter had planted and nurtured. The vegetables were attacked by a virus and worms. They died. I failed. My daughter, Nikiko, said not to worry: "We have other plants!"

"In our fields, we expect to fail. I do not protect my family from errors or problems. When we are sheltered from failure, we become paralyzed as if we don't have vision. We will always have more failures than successes, but that's why we practice. It may require two or three seasons, or it may take a lifetime. But that's what we have in common – we live with our failings *while* we practice in order to get it right."

I believed farming was a human profession filled with errors and mistakes – human failings. As agriculture adopted more and more technology, the pursuit of perfection changed. It was believed that by following a formula and precise methodology, the ideal peach or grape could be made. Once achieved, practice was no longer required; so-called perfection was mastered, then patented. Mystery solved.

"The dancer may, in a magical moment, feel perfection and cele-

moment. But the next day and the day after that, you return
lio to practice more, knowing the next performance may
very well not be perfect. Nor does it need to be. Dance is filled with
natural inconsistency and yet spontaneity. It is art — just as growing a
peach remains an art."

With modern farming methods, shoveling seemed to imply failure. After all, wasn't a weed an unwanted plant? Digging them meant
you failed at prevention, with a mistimed preemergent herbicide or
erratic disking. Shoveling was a penalty, a backbreaking reminder —
unless you accepted weeds as part of a natural farm and shoveling as
part of your seasonal routine.

"I celebrate the flavor of peach with a shovel in my hands, my
work, my calling, my practice. Dance celebrates the joy of movement
and the power of physical expression. I cannot grow without the
human element; dance honors the flavor of human motion. Farmers
with practice would make good dancers, and dancers with time
would understand the feel and meaning of my shovel."

I continued thinking of dance long after the conference. The story
of movement. Physical expression. Exploration of theme with my legs
and arms and torso. Secretly, in the middle of a workday, privately,
with no one around except my farm dogs, I'll raise my arms and spin
and whirl. Sometimes I lead with my shovel, skipping over the irrigation furrows, splashing in the water — da-dah! I dip under a low
branch or swing beneath the vineyard wire.

Why?

For a moment I'm transported and the fields become my stage,
the vines and peaches (and now barking and curious dogs) my audience. A farmer's jig, a rural village folk dance, a Japanese *obon*. A slow
dance with shovel. Later I'll return and see my erratic footprints in
the mud and the irregular strides in the earth. The loose dirt is scuffed
where I kicked the surface or landed hard. I can see motion and life.

Farm Tools

HATS

I use an old straw hat in the fields. In the summer our heat beats upon the land and stays with us until late in the day. We need wide brims to protect us. Straw works because it breathes, allowing air to circulate and heat to escape. I also like its crushability – I have sat, stepped, and leaned on my hats. Though the fibers snap and the hats are never quite the same again, they bounce back for more use. I insist on using one of these old straw hats when on the orchard tractor, because they work the best with low-hanging branches. As I duck under a limb, I can feel the slender scaffolds and leaves brush against the straw and I hunker even lower so my skull avoids injury. Isn't this why some animals have whiskers, a natural sensory warning mechanism signaling a tight and narrow passage or danger? Even in close calls when I'm driving too fast and a hidden, jagged stub from a limb juts out like a dagger, my hat provides me with a half-second warning. I jerk my head down and receive just a bruise and bump as the stump shreds the straw instead of tearing my scalp.

Eventually hats wear out. A favorite old one is beaten and weary, though I try to keep it employed for months more. Dusty, dirty, sweat-stained, crushed, lost and found. Dad used to buy only good-quality hats; the straw was thick and the weave tight, and he kept them for years. I'm either a duped consumer in the new throwaway economy or still a young farmer, a bit careless and impatient, whose best hats did not survive getting run over by a tractor. As seasons change, I try to start over with the new.

Baseball caps are a recent phenomenon out on the farm, and I doubt they will last long. They offer little coverage from the sun and weather, and a generation may soon discover they are scarred with skin disease and cancers. The caps, though, make for free advertising as we are bombarded with the continuing industrial-ization of farming. Ford, John Deere, and a number of chemical companies and equipment manufacturers promote brand names on the heads of farmers. I have often advocated that these compa-nies should pay us farmers for free advertising. I can often tell what crop a farmer grows by his hat. Various pesticides are regis-tered for row crops, others for grapes, still others for orchards — all proudly promoted on the head of the accompanying end user. Used to be farmers were a "Deere" family or "Massey" farm, but most have lost any loyalty. After all, Ford no longer even makes a tractor, despite its founder's vow "So long as Fords are made, we will make a tractor."

Hats are regional by nature. Wool for the hard winters of the Northeast and Midwest, stocking caps for freezing morning chores of milking. Straw for summer but wide brims for the Southeast and its long days, the light weave allowing air to circulate through the mate-rial yet casting a wide shadow of protection from the burning sun. I've seen sweat stains all around Southeastern farmers' hats, a white, salty reminder of their terrible humid and draining workdays. Mountain states must have stiffer hats to face the wind, the Northwest has to repel rain, and the hats of New England farmers must be as varied as their weather.

Older farm hats tell a story of a rich history. In the past, men often used felt hats. They would retain their shape even after a necessary slap on a horse's or mule's rump, urging it along down a trail or row. I imagine the view of the horizon as a farmer plowed his field, a harsh world from just below the brim, squinting at the many more passes required to finish a field or the miles and miles to travel before he reached his destination. A cold hard stare partially obscured by a floppy hat. There's more to hats, but I have trouble getting to their stories. Most farmers have their public hats, the ones they wear into town for a meeting or delivery. On the other hand, we also have our private hats, the favorite stained, dusty, worn, cracking, splitting hats that we sometimes try to sew and tape to repair. I have seen farmers duct-tape a battered farm hat. I have heard a threatening farm wife toss an old hat, delivering an ultimatum – "The hat or me" – and then, often quietly, the farmer retrieves it from a trash can. I have kept old hats even after accepting their death, hanging them on hooks in the mud room, believing for some reason they just might be used again. Instead, they gather dust and fade.

The hats of farm women have differed. Scarves and bonnets made from old fabric covered their skin, protecting them. Japanese-American women, especially the first-generation Issei, sewed bonnets made from old rice sacks. The long side panels kept the sun off fair skin, and the wide and coarse front brims blocked glare as these women toiled with hidden faces. Bonnets were like an American quilt, materials from their wearers' cultures, patterns handed down from generations, a fabric of who and what they were – workers of the land.

I often see women in the fields with bandannas around their heads and covering much of their faces. Their scarves seem to guard their autonomy as if they don't want to be recognized. From shame? Or a fitting symbol of an invisible laborer?

I've visited homes of widows whose farmer husbands passed away years earlier. Hanging on a hook or tossed on a shelf, the farmer's hat still rests in its place. The farmer still at home.

WORK BOOTS

I abuse my old work boots. They get caked with mud in the winter and soaked with slimy ditchwater of summer. The leather surface has been punched and poked by dropped pruning shears, cut and sliced with chain saws and clumsy hands, scraped and bruised when feet shove, push, and finally kick crooked boxes into place. Their soles have at times become a battering ram when I need to jam and force a tractor hitch in place. Worn heels can be swung, transforming a foot into a miniature club in order to loosen a trailer latch.

On a farm, work boots find a home in the dirt. I rarely walk on smooth surfaces, so the bottoms of my boots wear unevenly. The outside edges of both heels go first, and often the front toe area cracks and splits. I can stamp my feet a hundred times and dust will always, *always* fly up – layers and layers of dirt deeply embedded in the leather as if collected over generations like an antique in an old farmhouse attic or barn, lying around and collecting dust.

After a long workday, I kick off my boots and feel a rush of blood to my sore feet. They seem to take a deep breath and relax. With a flick of one foot, then the other, the pair sail through the air, tossed a few feet away. I'm free of their shackles, for at this time of day they are synonymous with work. One will land upright, the other on its side. Laces remain tangled and hang limp; strips of mud drop from grooves on the soles. A strip of fine sandy loam earth falls from the edges and marks an outline of the shoe on the gray cement floor.

The pathetic pair sit alone in the middle of our farmhouse mud room, a transition space where I change from farm clothes to inside wear. A small cloud of dust hovers above the boots, hanging in the late-afternoon sunlight. For a brief moment I think of moving the shoes, slipping them beneath my chair or under a low table, out of the way, lined up at a right angle, prepared for another workday. But I leave my boots scattered on the floor — a symbol of frustration while challenging the chaos of nature? Or out of exhaustion following a tough, physical day? The next morning I find them in the exact spot I left them, like a faithful friend I have taken for granted.

Each workday, I wear my boots for hours and hours. They gradually mold themselves to the unique shape of my feet, stretching over muscles, curling over each bone. Sometimes they stay on for twelve to fourteen hours, other times I whip them on and off throughout the day, giving my aching feet air. When I slip them off at midday, though, it's usually more than just a break for a few minutes. No shoes means rest.

I don't take care of my boots like others. I'm not the hunter who faithfully polishes his foot companions with each outing. My boots haven't experienced the attention of hikers or climbers who treat their shoes with reverence, knowing that their feet will propel them through most trails and passes. I have never polished a pair of work boots. Should they break, the laces are replaced only after an intricate series of knots have been tied in an attempt to extend their useful life. These are cheap working boots.

But I rarely throw away my old pairs. I collect them, shoving one pair under a table or chair, then, a year later after a new purchase, cramming another beneath a shelf. Initially I lined them up, like reinforcements in case my new pair was wounded and I needed an immediate replacement. As the old leather grew stiff from lack of use, spiders spun their webs in the openings and a blanket of dust dulled the shiny brass eyes. The line is disturbed only when I add another worn pair, pushing the row to the right as I retire the old.

Why do I keep these old shoes? They're like old friends, and it's hard to say goodbye, and I seriously believe I could find a use for them again. Perhaps I may assign specialized roles to each — one pair for rain and mud, another for days on the tractor when worn soles don't matter, a final stretched pair for cold, cold days when I wear double socks. Of course, this never happens. Once I strap on the current work boots, it is generally a waste of time to change them. Our conditions in central California don't warrant drastic equipment changes; it doesn't rain that much, and cold fronts offer an excuse to stay inside. Yet for years I have faithfully stored the old work boots. They have become a collection of my past, a timeline of my years on the farm. Perhaps they are like badges of honor, symbols of past battle campaigns.

The modern-day equivalent may be the old computers collecting in closets and basements. Rarely can we recycle these machines. We buy new ones to replace them and are not quite sure what to do with the old. We've spent hours at these keyboards and in front of these monitors. We created projects, expressed ourselves in memos, and wrote letters, e-mail, and journals on their screens. They were working one day and switched off the next. We have pangs about simply tossing them away.

There is a difference, though. Technology advances so fast that our old computers become obsolete before they get old. Computers change so quickly that we don't always develop relationships with our machines. Unlike industrial-age machines — a manual typewriter, a push mower, a gas-powered pump — they don't require us to learn personal tricks like a whack on the side of the gearbox to get the wheels turning when we first start up. Instead, we throw away and buy new when things break.

Work boots, on the other hand, wear out. The old boots carry the scent of our sweat, a distinctly human smell. They adapt to the shape of our feet. I have never lent or borrowed work boots. We wear new boots for months, breaking them in and allowing time for a relationship to evolve. Work boots age slowly. Computers are fast-food technology.

I discovered a pair of work boots from twenty years ago and

tossed them and a few others into a trash barrel. I cleaned out a shelf piled with old shoes, realizing I will never use them again. My old boots were a timeline of my past, but since my feet haven't changed since I was a teenager, my line of boots all looked similar. It was I that had changed. Staring at decades of old boots made me feel old. Tossing these relics of the past didn't make me feel younger or older. These old work boots were not obsolete but simply worn out. I let go of some of the past and said goodbye.

As I age and farm, I am wearing out like my old boots. But I hope my work and life won't become obsolete. I have finally learned that when boots grow old, I can gratefully get a new pair, and for the coming year, this new companion and I can grow old together.

PLOWS

One winter I questioned Dad about weeds and wondered how he took care of them before herbicides. We began a series of long conversations with stories about old farm implements, and often "did lunch" at the kitchen table with detailed discussions about shovels and cutting tools. He mentioned "Bezzerides," which I had never heard of. He tried to describe it to me, but it was difficult — old farmers often just repeat the same basic details over and over in their descriptions.

"A cutting bar."

"A metal cutting bar."

"Goes round the trunk."

"You know, a cutting bar around a trunk."

Dad had a habit of snapping his fingers when he couldn't think of the right word. Then he repeated the description, flicking his hand back and forth in front of him as if slapping some imaginary face. I watched clueless. Sometimes you wonder about oral histories and how recollections blur, facts and reality distort, unsolicited and rambling memories confuse what was with what was hoped for.

Dad tried creatively to combine the phrases. "A metal cutting bar." He added an ethnic reference as if that would help. "Bezzer-Ridi. I think they're Italian, two brothers just the other side of Reedley" – a neighboring town. I shrugged and nodded in agreement. I knew of another piece of equipment quite well – a few French plows were still used in the fields, especially the vineyards. I was sure the Italian immigrants to California's farmlands had their equipment too; weeds were very democratic, and the French did not have a monopoly on solutions.

A few days later I spied Dad going through our junk pile searching for something tossed away years ago. He pulled a flat, pale blue strip of metal from a pallet stacked with a jumbled heap. It looked like the leaf spring of a car or truck, one of the series of tempered steel slats that are mounted over the rear axle and provide flexible support. Automobile leaf springs are gracefully curved so when you go over a bump, they flatten to absorb the shock.

At lunch the next day he brought it out and proudly showed it to me. "This is a Bezzer-Ridi." He waved it in front of his face, as if the metal slapped some imaginary foe.

He also laid a yellowed, torn piece of paper on the kitchen table, an old advertisement about a "modern spring hoe" from Bezzerides Brothers. The illustration showed that the blade was actually mounted upright like a flat arm to shove dirt and weeds out of the way. The "spring action" allowed it to rub against a tree or vine trunk, bend back, then whip forward into cutting position. It slapped weeds.

Apparently the company was once very well known in the area, founded by two Italian immigrant brothers who farmed themselves and invented new weeding methods. But with the advent of herbicides, their sales plummeted and their invention almost disappeared. Long ago, Dad had a set of Bezzerides blades, and they worked well until he snapped one (that confession took another two weeks of lunches). He tried to weld the tempered metal back together, unsuccessfully. He offered sage advice: "Don't back up with them. They only bend in one direction."

Over that winter and spring, Dad and I talked about slow yet simple mechanical methods to control weeds. Herbicides had displaced these tools, and experience and knowledge grew obsolete. But Dad seemed to come alive with my questions, and a series of other "cutting bars" were produced from our junk pile. I'd experiment with them, uncertain how they were even mounted on a plow or disk. We took turns on the tractor, the other watching the tool at work, then making adjustments.

The cutting bar swam back and forth in the earth, tapped a trunk and then snapped into place until the next vine. I enjoyed hearing the rhythm of the work. The Bezzerides worked great, but only when the weeds weren't too large. We missed that window of opportunity because our weeds had already grown to be a foot high in bad spots, a difficult lesson that first year. We fought the big weeds all summer and would have to wait until the next winter for another round of lunches in order to refine our technique.

ELECTRIC MOTORS

We cool our farmhouse with a "swamp cooler." It works as follows. Three damp fiber pads surround a spinning fan. Air is sucked through those moist pads and cooled before blowing into our home. It works well in the dry, dry heat of the San Joaquin Valley unless the temperature reaches one hundred degrees with seventy-degree nights. Then the air remains hot and all the cooler accomplishes is to make your skin feel sticky. Our house then feels like Florida.

I enjoy the economy and simplicity of the machine. I also recall memories blowing through swamp coolers. As a kid, I shoved and pushed my brother and sister away and claimed the sweet spot directly in front of the blowing air from our window-mounted unit. Actually it didn't matter. We were small enough to fit if we squeezed close together, but the challenge was to do so without touching each other. Later we put paper streamers on the front screens (we had seen this at a hardware store) and let them tickle our bare stomachs, giggling and testing to see who had the most tolerance. Sometimes I pushed a small lounge chair next to the cooler and lay under the magical air, the wind rushing over my face, comforting a child's hot body and feet.

I can picture my dad standing in front of the three (or four with the deluxe version) faded white plastic vents with angled slats that shot the air in different directions. He'd take them off so all the air would blow directly on him. In the evening, he'd come in from the fields, practically take a sponge bath while washing up at the outside sink near the back door, come in with water dripping off his hair and chest, and stand for ten minutes in front of the swamp cooler. He looked as content as a dog lying in an irrigation furrow, eyes closed, a slight grin on his face, the warmth of the hundred-degree summer workday still radiating from his skin. Mom would tell him he'd get a chill if he did it too long. He'd pretend not to hear and just let that cool wind stream across his body, blowing the day's work and troubles away and welcoming the evening with family.

Later when my folks got central air conditioning, we insulated ourselves from the dry heat outside and hid within the confines of our house. I lost those views of Dad and no longer had to battle my siblings and squeeze in front of a cooler. When I moved into our present farmhouse, I was happy to discover it only had a swamp cooler.

But now our cooler motor had burned out – the water tube had come loose and showered the electric coil with spray and shorted the windings. We could smell the burning wires in the house. I pulled off the motor and went to visit Dad. Old farmers always have about a dozen electric motors lying around. Some motors are mounted on antiquated equipment – a conveyor belt we once used to sort peaches or a fruit sizer that was still sitting in the same place it was twenty-five years ago when we stopped packing our own fruit in the shed. When Dad tested one of these ancient machines to see if it worked, he found the cracked dull-gray electric cord, coated with years of dust, and instinctively turned to his left, leaned over to where an outlet lay hidden behind boxes, and repeated the ritual he must have followed for years to mark the beginning of another summer workday. It worked: the motor hummed with life.

Some motors sit in obsolete swamp coolers, displaced by newer ones, relegated to an obscure corner in the shed and later cannibalized for parts. Other motors are shoved in boxes and stacked on shelves or set on the floor under a workbench. Another lies outside in the rain, the shaft and windings forever frozen in place. Two are still in their boxes but never used.

Electric motors are from my Dad's generation of the 1950s. Electricity reigned on the farm, and small family farmers became a hungry audience for budding inventors and engineers who captured the energy of motors to increase productivity, power new machines, and catapult a farming operation into a new era.

In my old barn I occasionally uncover the remnants of the era before electricity and tractors. The farmer before us had collected a dozen harness "trees," used with a team of work animals to keep the ropes and straps from getting tangled. The metal rings were rusting, and beneath decades of dust I could see the hard wood was worn smooth. The trees hung on barn walls, lay tossed on a workbench under piles of other abandoned equipment, and in one case propped up a low board that held a row of old tires against the wall. Each tree had a history of its own and a specific job — the longer ones for more power, the shorter ones for maneuverability. I imagine different teams preferred one over another. Did mules like the fat ones while horses the skinny, slender ones? If you had to work at a fast clip, did the lightweight one work best? For slow, prodding work like plowing, did you use the one with thick rings? I can only guess.

Dad walked around my burned-out motor, curiously studying it while searching his memories. "Let's see," he said. He crouched low to read a blackened identification number. "Could have come off the old sizer belt." Once we used to pick and pack all our own fruit for shipment. Dad made most of the equipment to automate the process — a sizer to speed up the packing, a belt to deliver the fruit with less handling, a "defuzzer" that had a series of roller brushes that gently wiped the fuzz off peaches. Hundreds of small family farms operated like this, miniature Henry Ford factories with assembly lines revved up each summer, using family labor and all powered by electric motors.

"Before the sizer," Dad added, scratching his chin. "May have been part of a cooler from the old house. 'Bout forty years old." He slipped into his past with each memory of another motor. "But a third horse-power, that doesn't fit anything." He shook his head. "Maybe picked it up at an auction, but didn't have much use for only a third unless it just operated a fan. . . ."

We shared a silent moment. I knew not to say anything.

"Or remember the plumber neighbor John? He may have left that with us when he died. . . ."

I motioned to the new motors still in their boxes. Dad thought for a moment, then his memory connected with a story. "Let's see . . . it was when Uncle had his car accident. . . . Planning on building a new fruit-sorting table that spring, but had to take care of Uncle's place instead. Never got around to making that one. . . ."

We opened one of the dusty boxes. Dad stroked the smooth gray housing of an unused one-half-horsepower motor. I wondered what sorting table he had in mind and what other unfinished business lay scattered in the shed. The box read "Dayton Electric Motor Company." Could they still be in business? I imagined tens of thousands of Dayton motors powering our nation a generation ago, driving fruit sorters and conveyor belts and hopes and dreams of inventors and family businessmen. Electricity paired with motors were magic. They could transform a small family operation into something. Each new innovation of Dad's did increase production, leaping from a few hundred boxes a day to a peak of a thousand.

Our old barn turned into a miniature factory each summer. We all had our jobs and motors to help us. Each workday, they whined at our feet, buzzed by our heads, and droned in the background. We'd turn on the radio as loud as we could and could barely hear the top twenty hits of the week over the hum of the captured energy. But in the evening, one by one I'd shut off each motor, shutting it down for an evening. They radiated heat until the night arrived, and then they cooled in the dark, resting before the next morning, when Dad would flip switches, the motors would turn, and a new day would begin.

I took the new Dayton motor and mounted it in my old swamp cooler. I had to redrill the bracket to fit the new motor. Years ago I had learned that nothing with this old farmhouse was standardized. Brackets, braces, rough-cut joists, pipes, electric conduit reflect the time they were installed, individualized to an era and the specific craftsman. Their work does not always fit well with today.

The new Dayton motor would cool me and my family at night. When this motor burns out, I'll tell my children the story of where it came from and all about the belts and sizer and the hum of the motors, about the plans for a new fruit sorter that was never to be, and about their grandfather the inventor.

Slow Pickup
Trucks

I HAVE A slow truck. Actually, we have three. A '74 Ford Courier sits in my dad's yard where I parked it when I got a new truck over a decade ago; another truck rests in my Dad's shed, a '72 Chevy that he last drove in 1989. He couldn't start it that winter and simply left it there. His shed is not a rustic old wooden barn leaning slightly to one side with a rusting basketball hoop dangling from the side near the old hayloft. His truck sits on a cement floor under an enclosed aluminum structure that houses the family car and the young '81 Chevy El Camino. Dad still occasionally wipes the dust from the windshield of his parked Chevy.

My latest "new" truck is a '91 Mitsubishi Mighty Max. They don't make them anymore, but I can still get parts, so Dad and I consider the vehicle new. Besides, this is a farm truck with only eight thousand original miles in ten years, mainly back and forth for a quarter of a mile between my house and Dad's or up and down the farm's vineyard and orchard avenues. Once I calculated that would equal about fifteen hundred round trips to their house per year or about four times a day — and I thought, "Yeah, that might be right." Then Marcy reminded me of the miles lost because I do a lot of backing up – when I check water and drive along the ends of the rows, monitoring where the irrigation water has or has not yet reached, backing up to confirm which rows need adjusting or surmising a furrow must have broken in order for an odd row to have that much water; or going in reverse because it's too much trouble to turn around. I've mastered the art of driving in reverse, fairly quickly, trusting only my mirrors.

I don't remember ever going fast in this farm truck. It's a five-speed, and I can count on one hand the number of times I've been able to get the speed up to use fourth gear and don't ever remember using fifth. Mostly I drive slow down rough dirt roads with potholes whose location I've memorized so I don't even notice them anymore. I unconsciously swerve and can feel my buttocks automatically tighten, bracing for the jolt. I coast in neutral a lot in this truck – that's the proper speed of good farming, slow enough to monitor the world beyond the windshield while still making progress. I imagine it's the pace of farmers riding a horse or walking a lot, but I like the feel of coasting and often shut off the engine and glide for a while, covering ground and listening to the sound of dirt crunching beneath the tires. I pause, believing my work is just too important to rush, convincing myself there's status to going slow. When I finally come to a rolling stop, I can't help but enjoy the silence, a moment of fullness at the end of a chant or meditation or prayer.

The instrument panel has few dials, no radio, no clock (which would have stopped working long ago anyway), no air conditioning, and the heater sort of works but when I turn on the fan, a blast of dust comes out of the vents first. The gauges are covered with dust, and the only one I can read is the one for gas. Surprisingly, only a few cracks blemish the bench seat, although the driver's side has a familiar sag molded exactly to my bottom. The body is filled with numerous dents and dings. Most of them I can account for, like the time I nudged a post while backing up with my tailgate down or the scrape in a door when I moved the truck ten feet and didn't notice a low pile of props, or the thousands of nicks in the bed where I tossed a shovel. This vehicle has aged with me.

My truck may never be considered a classic. The doors don't have a heavy, solid feel; I have never stroked the sleek metal. The early nineties are much too recent to inspire a nostalgic sentiment – even the seventies seem too close in my rearview mirror. Certainly a truck with the Mitsubishi emblem will never be considered an American beauty, but even my old Ford Courier was an import, and Dad's Chevy, while some claim to have come from the end of a golden era when it seemed everything good in life was about American cars, has little intrinsic meaning for us because these were all work trucks first and forever.

The best thing about my truck is that it continues to go slow. It wants to mosey along, take its time, more comfortable in first gear instead of overdrive. I tease myself with the thought I'm much too important to rush. "No sweat," I say out loud, grinning. Every time I chug along, for a few minutes the hardworking farmer rests; I can show the world I'm rich, not with money or respect but with time.

Touch Method and
Niki's Clutch

NIKIKO WAS about the right age at ten. She could reach the tractor clutch but had trouble pushing it down unless she stood up and used all her weight. I remember driving when I was much younger, part of a revisionist memory of an older generation when everything we did was harder, heavier, slower or faster (depending on the chore), and of course better for our character. Or did Dad start me out younger because I was a boy?

Learning to drive the tractor is a farm kid's rite of passage as soon as you can reach the clutch. Dad cheated with Rod, my older brother, the oldest child, because he needed his son's help at an earlier age. Boxing raisins in the late 1950s required raisin rolls to be hauled out of the vineyards first – a slow, very slow wagon driving up and down each row, Dad walking faster than the tractor, stooping over and picking up each roll and handing it to Mom, who rode on the wagon, stacking them in a pyramid pile, hundreds of rolls on her wagon. Didn't want to waste a good worker as a driver, so Rod was fitted on the old Ford, lowest slow gear, just keep steering straight and stop at the end of the row so Dad could negotiate the turn. Dad taped wood blocks on the clutch so the frail little boy could reach the pedal and stop. I was pressed into service at a later age than Rod and didn't need the blocks – not because I was taller with longer legs, but because I was fat and had the weight to work the clutch.

Niki tried with all her might, but the tractor kept churning forward. She stood and pressed down, with no progress. A look of terror began to sweep across her face, eyes wide open, tight lips ready to explode in a cry for help. She jumped, and the pedal slipped down; only with all her might did it engage. A sense of relief as the tractor coasted to a stop.

"Do I have to do this every time we need to stop?" she pleaded.

"Yep."

Equipment not designed for women, especially a little girl. I talked with a group of farm safety experts from the Midwest. They were worried because more women were doing more and more of the farm work – a sign of harder times, a necessity as old farm husbands died – and were discovering tractors and harvesters were not designed for their bodies. Accidents happen. Was I rushing Niki?

Our tractors have changed. Niki could have easily handled an old Ford 8N, smaller with less horsepower, gas engine, ancient clutches with lots of play. Best way to stop was to yank the gearshift into neutral, and if you timed it right, you could do it without even clutching. Later, we bought diesel tractors, mostly big Masseys with a few wide Fords, all more powerful, with sixty horses, solid frames, and stiff and heavy clutches. You had to commit to stopping.

A few lessons later and Niki had figured out a method – "stomp" was her mantra. Stopping and starting in first low didn't require much touch. I could stroll alongside the tractor, and easily climb on board should her stomping falter. Next we tried a faster speed, third or fourth gear, in which a clutch, once engaged, would jolt if she wasn't careful.

I explained to Nikiko: "Now we add touch."

Her short-lived confidence was crushed with the first whiplash start, then a slam forward and a sputter that killed the engine. I laughed, she glared. We started over with the same results. Half a dozen kills later the tractor battery was beginning to drain so I suggested that we stop. She was not happy. My words only seemed to annoy, because kids aren't allowed to fail much these days.

"It takes time. Touch takes time."

I remember watching both Niki and Kori as youngsters when they drew pictures of people. Big heads and large hands because that's how they saw the world and how it felt to them. My contribution — enlarge the feet, shrink the heads, and hope for bigger hearts. I want them to feel a lot — the "give" in peaches and the "chill of fog down to the bones." I want to teach them the inexact farmer science of measuring moisture in the land by squeezing a handful of dirt — if it sticks together there's enough to delay an irrigation, and if it falls apart it's time to schedule a drink. I like my friend's description — it's like shaking hands with the earth. Will my children be the last generation to know how to measure moisture in raisins by rolling them in their hand? It's like cooks who teach their children how to determine the consistency of dough by rubbing it between the thumb and fingers. Niki and Kori have grown up in a tactile island surrounded by a growing nontactile world. But that's not totally true. Perhaps everyone's sensibility about touch has changed; after all, fast food is to be eaten with our hands, though few of us think about that.

Nikiko knows the farm, while Korio is still learning. As she climbs back on the tractor, determined to learn this craft, I can't help but stare off into the unknown. I can't see their children learning about tractor clutches. It's not that they won't — I just can't see it.

When Nikiko graduated from eighth grade, she marched in as part of a long line of excited and beaming students. Her junior high was large for a rural school, over three hundred in her class alone. Marcy and I were proud. She was in the top of her class and received

awards and a savings bond for college. She was ready to conquer high
school and full of confidence about her future.

We were surrounded by ecstatic families, but were proud in a different way. Others were cheering, yelling, screaming, crowding to
take photographs, borderline rude because after their child marched
in front of us, they no longer paid attention. The audience filled with
chatter about graduation parties, huge affairs with dozens of guests
and a vanload of gifts. I was stunned to see in the parking lot a number of rented limousines. For an eighth-grade graduation.

I felt bad about a small gathering we planned to commemorate
Niki's graduation, one of many steps in her educational journey. I
hoped she would not feel too disappointed comparing her celebration
to those of some of her classmates. Even before the final diploma was
given, the crowds began to filter toward exits and festivities, and the
limos filled and made their way to the streets. I chatted with some
teachers, already preparing for a deserved summer vacation. When I
asked if limos were part of every graduation, Elaine, an astute veteran
teacher, rolled her eyes and explained, "I can't blame them – they celebrate because for thirty percent of this class, this will be their last
graduation. This is as good as it gets." She sighed, a long year of teaching over.

As Niki prepared to test her clutching skills one more time, I
imagined the sound of a scratchy graduation march playing over a
loudspeaker and saw her as part of a long procession that didn't end
with the eighth grade. I was sure she would continue to do well. Yet
with each success, I saw her pulled away from the farm, especially as
a young woman meeting challenges and seeking new opportunities.
And Korio would soon learn how to drive a tractor, but would his
success pull him from these fields too?

I imagined some farm kids staying here, but only a few. I envisioned many of those other eighth-graders from Nikiko's graduation
settling in the community, not as farmers or farm workers but work-

ing in support industries — the box company, packing sheds, trucking outfits. Some will become mechanics I trust and befriend.

Finally, the tractor jolting smoothened as Niki began to understand the clutch. I explained about the sweet spot of this tractor, the point where the engine contacts and connects with the transmission and the power is transferred to the wheels. "A fine line separates standing still and moving forward. Feel the moment when it engages." Like all farm kids, she struggled and eventually improved, albeit with a lot of jerking and engine stalling. "Play with it," I added, "and you'll find the touch."

Savory

Things Worth Saving

Old peach varieties. Sun-dried raisins. Slow trucks. Junk piles. Farm dogs and cats. Chickens. Barns with barn owls. Porches. Straw hats. Rice-sack bonnets. Fruit labels. Sweat boxes. Worn shovels. Dump rakes. Manure spreaders. Tractors your kids learned to drive on. French plows. Wildflowers. Mud. Sweating. Farm kids. Old farmers. Baachan's/Grandma's garden. Extended family. The art of — grafting; squeezing peaches; feeling raisins for moisture; irrigating uphill. Blossoms. Scent of grape bloom. Farm walks. Memories. "Next years!" Silence.

Headstone
with Blossoms

AN ELDERLY Nisei man sitting next to Marcy at church suddenly asked, "Did you know your peach is on a headstone?"

Bewildered, Marcy shrugged and shook her head no.

The man added, "In a cemetery just south of San Francisco."

Marcy was then reminded of a woman who years ago came to visit our farm and was so moved that she asked that a peach be engraved on her headstone. I recall a visit from two older Nisei sisters. We were in the final stages of a long harvest, and I was still frantically sprinting to get a final load in. Marcy greeted the visitors, and they took a slow walk into the orchards and found a single peach still hanging in the tree. Alice, one of the sisters, had read about our peaches and family farm and wanted to see them for herself. Five months later she was dead from cancer.

Finding the cemetery is proving to be difficult. Even with a map and directions, the freeway exit is confusing, and I have to zigzag my way through smaller streets and oddly angled intersections. I'm in a rush too, squeezing in a visit between a business meeting and a presentation in the San Francisco area. I can see a larger cemetery across the street, and pulling to the side to reexamine the map, I look up and discover a smaller and strangely familiar cemetery. An entrance marked with a tiny sign with fading paint: Japanese Cemetery, Japanese Benevolent Society of California.

A cold breeze greets me on this Sunday morning. Spring is breaking winter's crust back on the farm; here, the sun shines but the crisp chill lingers. I hold my arms close to me. Past a stone pillar with Japanese *kanji*, I walk into another world. As in Japan, the headstones are narrow vertical columns, tightly clustered so they don't take much space. The granite stones are mixed ages, and in the center sits a single large monument with the Japanese *kanji* for *sho kon*, "inviting spirit." A lone elderly Japanese-American man with a dark sweater is strolling down one lane. He slouches slightly, leaning forward, holding his hands behind him. A morning visit with his friends?

The older section of headstones is to the right. I walk past a large pillar dedicated to George Shima, the Potato King, pioneer farmer. Nearby stands a monument for Buddhist ministers. I recognize the name Hogen Fujimoto, a teacher and a friend who long ago published one of my first writings in a Buddhist newspaper. I stop to read the list: these names are from my past. The cemetery seems a little larger than a football field, enclosed by a wire fence. Behind me cars roll past; on the other three sides are the backs of single-story buildings, probably businesses, and homes with families slowly waking up.

The new half of the cemetery is largely empty. An asphalt path winds through the open grass field. Only a few markers stand along the way; most are rectangular beds filled with white stone and a simple black granite headstone rising above a final resting place. I was told the peach was on the new side.

I walk along the fence, circling the edges, gradually moving closer to one grave near the center. Bouquets of flowers, yellow and scarlet mums with carnations, sit upright on each side of the headstone. As I turn to face the grave, I am stunned to see the name Matsumoto etched in large letters – no relation, with a slightly different spelling, yet it sends a shiver, as if my family is part of this place. Below the family name is carved "Alice Abe," who had once visited our farm.

Above, on the top third, is the peach. A single peach hanging on a branch, a cluster of leaves gently protecting it and keeping it company. To etch the peach, the black exterior facing had been chiseled away, exposing a gray-white interior. The polished surface contrasts with the rough interior texture of the fruit's skin; the peach and leaves are engraved in the stone as in a woodcut.

Were I in Fresno, Modesto, or Marysville, where farming and orchards thrive, this would all be expected. A tribute to a farm family, the history on the land, a woman's legacy of working the earth. I had learned, though, that Alice was born and raised in San Francisco and never spent much time on a farm or out in the fields. After World War II, she married and lived in the Minneapolis area. Only later she had read one of my stories and wanted to learn more about our farm, our family, and, of course, the peach.

Her husband, Yo, writes in a letter: "Marcy directed us to the orchard but said the peaches had all been harvested. However Alice did manage to find one peach hanging from a low branch. I remember her happiness that day in finding that peach. Shortly after we returned home, she started chemotherapy, which was not successful. Her final wishes were for a peach to be part of her headstone. I remember a phrase you wrote that 'eating a Sun Crest peach automatically triggers a smile and a rush of summer memories.' That is what the peach engraved on Alice's headstone does for me."

A peach from our family farm. A peach from my stories. Now part of another story. That peach was no longer mine. This is why I farm and why I write. Too easy to get caught up in production and sales, markets and marketing, demand and reviews, profits and losses. I sometimes forget why I love this work — to grow a peach or grape or raisin and hope there's a story shared in a simple, honest way. And to comfort a woman in her final days.

It's more than just the art of self-expression: I farm and write with the spirit of the humanities. Human — i — ties. I'm one of the many — the "i" between human ties. The art of human ties. The sound of a ripe peach. The taste of a single raisin. The aroma of a farm father. The angle sunlight on the flesh of a ripening peach. The feel of old work boots. A spirit of connection and the power of stories.

Quiet stories that don't shout out for attention, subtle flavors a final generation knows from memory. A type of folk art or craft that may not always fit in a museum or catalog or grant application. No different from losing a song or a dance, the loss of one more farmer and the sale of another family farm unfolds like a tragedy, a drama with no one in line to fill a role. We would be foolish if we stopped teaching the rhythm of a growing peach or the feel of sweat and dust as we rope and prop old trees. Tastes and aromas that journey beyond the written word and defy boundaries, art that touches in the immediate.

I leave some peach blossoms at the headstone. Five short branches are filled with dozens of flowers at different stages — some not quite ready to burst open, others older, with the light pink color beginning to fade and edges curling brown. I pause and stare at the engraved peach and grin, then step back and see my reflection in the dark smooth granite. As I turn away, the delicate blossom petals scatter in the breeze.

Things Slow

Farming. Pruning. Irrigating. Suckering. Girdling. Touching. Hand saws. Hand weeding. Hand picking. Grape knives. Raisin trays. Ladders. Buckets. Vineyard wagons. Smelling. Organic fertilizers. Natural predators. Blossoms. Raisins in September. Your father's smell. Listening. Porch swings. Screen doors. Sash windows. Summer afternoons. Falling leaves. Bearing trees. Fog. Seeing. Grandfathers. Fathers. Grandmothers. Mothers. The farmer's wife. Kids growing up on a farm. Kids leaving the farm. A field getting old. Tasting. A single raisin. Childhood memories of what peaches used to be like. Hard sweat. Exhaustion. Tired. Calloused hands. Waiting. Dying.

Acknowledgments

❧

Writing and farming humble me daily. Fortunately, I am surrounded by family and friends who continue to believe in my voice. I am grateful to them and hope my stories return the many gifts of support.

Also I am grateful to the many people and audiences who have listened and allowed me to share my thoughts and words — each exchange helps me find my voice. Thanks to Copia, the American Center for Wine, Food and the Arts, to Yale University and its Program in Agrarian Studies, to Dance USA, to Chamber Music of America, to Agriculture in the Classroom, and to the numerous educational organizations and teacher conferences at which I have presented.

Acknowledgment is due to the following publications where some of these essays appeared in a slightly different form: *San Francisco Magazine*, *Contra Costa Times*, and the Patagonia catalog.

Much thanks to Joyce Mills and her eye for the valley and her skill at art — her drawings complement my words like a fine wine. Also thanks to Scott Willson and Patagonia for the striking cover photography that captures the perfection of a peach.

Special thanks to Elizabeth Wales and Alane Mason for their many skills and continuing support. Also to Adrienne Reed and Stefanie Diaz — their assistance is not forgotten. My work would not be complete without help. Thanks for believing.